Critical Insights From Government Projects

CHARTRIDGE BOOKS OXFORD

Thoughts With Impact Series

The purpose of the *Thoughts With Impact Series* is to act as a knowledge repository covering the key fields of government practice. The Series also aims to: support the mechanics of knowledge-building in the government sector and public administration; contribute to understanding the building blocks; and identify ways to move forwards.

The Series recognises that it is important that to analyse government practices from a systems theory perspective. We need to view government and public administration from a broader perspective and not depend on the events and circumstances in our own environments. Such broad thinking and analysis should guide and help us focus on outcomes rather than reacting to events as has always been done in the past.

Titles in the *Thoughts With Impact Series* include:

- *Critical Insights From Government Projects (number 1 in the Series)*

- *Critical Thoughts From a Government Perspective (number 2 in the Series)*

- *Critical Insights From a Practitioner Mindset (number 3 in the Series)*

- *Critical Thoughts From a Government Mindset (number 4 in the Series)*

Critical Insights From Government Projects

DR. ALI M. AL-KHOURI

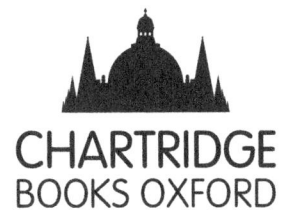

CHARTRIDGE
BOOKS OXFORD

Chartridge Books Oxford
Hexagon House
Avenue 4
Station Lane
Witney
Oxford OX28 4BN, UK
Tel: +44 (0) 1865 598888
Email: editorial@chartridgebooksoxford.com
Website: www.chartridgebooksoxford.com

Published in 2013 by Chartridge Books Oxford

ISBN print: 978-1-909287-51-8
ISBN digital (pdf): 978-1-909287-52-5
ISBN digital book (epub): 978-1-909287-53-2
ISBN digital book (mobi): 978-1-909287-54-9

Typeset by Domex, India
Printed in the UK and USA

Contents

Foreword

His Highness Sheikh Saif Bin Zayed Al Nahyan

The Minister of Interior and Deputy Prime Minister of UAE

His Highness Sheikh Saif Bin Zayed Al Nahyan

I am delighted and proud to write a foreword to this new Series of seminal books written by Dr. Ali M. Al-Khouri.

Governments the world over have been directing major efforts to drive their activities and services to their citizens. There have been and continue to be many national level initiatives and several programs that spanned years of implementation. E-Government initiatives have been launched to take the Government to the citizens' doorsteps. Technology implementation has thus become central to Smart Governance. Dr. Ali Al-Khouri has meticulously captured our unique journey so far in the science of Identity Management and Technology Implementation in his books. The reader will be delighted to see actual implementation references of various technologies in these books which are presented in the *Thoughts With Impact Series*.

The *Thoughts With Impact Series* comprises four current titles:

- *Critical Insights From Government Projects*
- *Critical Thoughts From a Government Perspective*
- *Critical Insights From a Practitioner Mindset*
- *Critical Thoughts From a Government Mindset*

The books in this Series are unique in that they are written from the perspective of a member state of the Gulf Cooperation Council (GCC) and cover highly topical subjects which all member states of the GCC are actively investigating.

Dr. Al-Khouri brings to his authorship a unique academic and executive knowledge that is based on the tenets of good governance, which undoubtedly is the bedrock of administrative excellence evidenced in the qualitative standards of living of UAE nationals.

One sees the effective and efficient, consensus-oriented, accountable, transparent and responsive characteristics of good governance, ably brought out in the Series: in the futuristic blueprinting of smart-technology based infrastructure, the systemic insistence on culturally relevant best practices, and the high level of integrity maintained in policy decisions of the government.

What strikes one most about the Series are the candid insights into what may not have worked in the process of policy implementation – insights that are the hallmark of a peer-reviewed scientific discourse that attempts to capture all the stakeholder considerations involved in the running of a government.

Starkly, the highest consideration that drives the author's involvement in the projects outlined in the Series is ensuring policy makers and implementers do not lose sight of the cultural, social and ethnic bases of technology adaptation while leading the path of global excellence in the rapidly evolving geo-polity of nations.

Dr. Al-Khouri details with technical alacrity what is perhaps the most ambitious of the UAE government's undertakings – the Emirates Identity project that exemplifies the successful working of government systems for a people; and the Iris comparison data of the UAE government's immigration department, which makes UAE a forerunner in putting into place an internationally recognizable security system that is mindful of contemporary socio-political realities.

The first book in the Series, *Critical Insights From Government Projects*, examines various projects in the government field and, more specifically, those in the GCC countries. Chapters in the book are grouped into four main categories: project management, projects evaluation, electronic services and technology implementations. These papers cover a variety of research topics in the government context. They represent experimental practices in the field of public sector management and the implementation of advanced technologies in a government setting.

Critical Thoughts From a Government Perspective, the second book in the Series, examines the broad picture concerning the realities of day-to-day issues faced by public sector managers. These papers address various areas of considerable importance for public sector leaders and so they have been grouped into four distinct categories: strategic management; organisational performance; e-government; and national identity.

Critical Insights Froma Practitioner Mindset assesses the thoughts, experiences and lessons learned from large systems implementations in the United Arab Emirates and GCC countries and reflections on similar implementations in other countries in the world. The book looks at four main areas of research: the new digital economy; e-government practices; identity and access; and, finally, identity systems implementation.

The latest book in the Series is *Critical Thoughts From a Government Mindset*. This book provides an analysis of three main categories of research: strategic management; e-government development and practices; and, finally, identity management.

All four books are well researched, thought provoking and inform the reader about the very latest developments in the field.

The extensive project-relevant academic literature reviews and citing of industry-oriented research material gives the Series the added touch of authenticity and credibility, offering a holistic view of even the most practical placement of systemic structures.

They are required reading for all researchers and practitioners worldwide especially those seized with the task of ensuring that the benefits of technology reach all levels of society.

I commend Dr. Ali M. Al-Khouri for his comprehensive, detailed and sagacious analysis. These four books provide an invaluable and significant contribution to research in the field.

Dr. Ali M. Al-Khouri is a distinguished academic researcher and practitioner, and an active researcher in the fields of advanced technology implementation in the government sector, how to reinvent governments and revolutionising public sector services and electronic business. He has published more than 50 research articles in various areas of applications in the past ten years. As the Director General (Under Secretary) of the Emirates Identity Authority, Dr. Al-Khouri helped establish this federal government organisation and rollout the national identity management infrastructure program in the United Arab Emirates. He has been involved in the UAE national identity card program since its early conceptual phases during his work with the Ministry of Interior. He has also been involved in many other strategic government initiatives in the past 22 years of senior governmental experience.

His Highness Sheikh Saif Bin Zayed Al Nahyan
The Minister of Interior and Deputy Prime Minister of UAE
June 2013

Author's note

This book was originally published, in 2008 and 2012, by the Emirates Identity Authority, Abu Dhabi, United Arab Emirates. Permission to republish this book is gratefully acknowledged.

Chapters in this book have previously been published elsewhere:

Chapter 1. Projects management in reality. Lessons from government projects. © 2012 *Business & Management Review*. Al-Khouri, A.M. (2012) 'Projects Management in Reality: Lessons from Government Projects', *Business & Management Review* 2 (4): 1–14.

This article first appeared in 2008: Al-Khouri, A.M. (2008) 'Why Projects Fail? The devil is in the detail', *Project Magazine* [Online]. Available at: *www.projectmagazine.com*.

Chapter 2. An innovative project management methodology. © 2007 *Warwick Engineering Conference*. Al-Khouri, A.M. (2007) 'A Methodology for Managing Large-Scale IT Projects', *Proceedings of Warwick Engineering Conference*, Warwick University, Warwick, United Kingdom: 1–6.

Chapter 3. The UAE national ID programme. A case study. © Waset.org, 2007. Al-Khouri, A.M. (2007) 'UAE National ID Programme Case Study', *International Journal of Social Sciences* 1 (2): 62–69.

Chapter 4. Using quality models to evaluate large IT systems. © Waset.org, 2007. Al-Khouri, A.M. (2007) 'Using Quality Models to Evaluate National ID systems: the Case of the UAE', *International Journal of Social Sciences* 1 (2): 117–130.

The content of this article was partially presented at an international conference: Al-Khouri, A.M. (2006) 'Using Quality Models to Evaluate Large IT Projects', *Proceedings of the World Academy of Science, Engineering and Technology* 21 Vienna, Austria.

Chapter 5. Electronic Government in the GCC Countries. Barriers and Solutions. © Waset.org, 2007. Al-Khouri, A.M. and Bal, J. (2007) 'Electronic Government in the GCC Countries'. *International Journal of Social Sciences* 1 (2): 83–98.

Chapter 6. Digital identities and the promise of a technology trio: PKI, smart cards and biometrics. © Science Publications 2007. Al-Khouri, A. M. and Bal, J. (2007) Digital identities and the promise of a technology trio: PKI, smart cards, and biometrics, *Journal of Computer Science* 3 (5): 361–367.

The content of this article was partially presented at a national conference: Al-Khouri, A.M. and Bal, J. (2005) 'Identity theft and the promise of a technology trio', *Proceedings of 3rd Safety & Security Conference*, Abu Dhabi, United Arab Emirates.

This chapter was quoted in: Summer 2007 Intelligence section, *MIT Sloan Management Review.*

Chapter 7. Reprinted from *Telematics and Informatics*, Volume 26, Issue 2, 'Iris recognition and the challenge of homeland and border control security in the UAE', pages 117–132, 2008, with permission from Elsevier.

Preface

This book represents a collection of published research articles in several international journals and magazines during 2007 and 2012. They cover various projects in the government field, and more specifically those in the GCC countries.

To allow better reading, the papers included in this book have been grouped into four research categories: project management, project evaluation, electronic services, and the implementation of technology. These papers offer a variety of researched topics in a government context. They deal with experimental practices in the field of public sector management and the implementation of advanced technologies in government sessions.

These papers can also be distinguished from studies available in the existing body of knowledge conducted in the Middle East. Research studies in this region are normally conducted by researchers who are very much interested in academic rigour, rather than its practicality. Also, very limited information is normally exposed and distributed about government projects which are by and large categorised as being classified, which makes existing research studies lack a fundamental understanding of issues that make up the bigger picture.

The research work in this book was written by senior government officials and practitioners. They bring forward key critical insights from several strategic government initiatives, from general management frameworks, and from imperative thoughts, reflections, and fundamental lessons learned. This should allow management to deepen their understanding of such projects and practices and better manage the associated risks.

In short, the intention of this work is to support the development efforts in organisations in the GCC countries and contribute to the advancement of the overall fields of research.

I hope that you will find this book immensely accessible and practical.

Dr. Ali M. Al-Khouri
2013

About the author

Dr. Ali M. Al-Khouri

Dr. Ali M. Al-Khouri heads the Emirates Identity Authority, a UAE federal organisation, as Director General. He received his Engineering Doctorate degree from Warwick University where his research focused on the management of strategic and large-scale projects in the government sector. He is a Certified Project Management Professional and a Chartered Fellow of the British Computer Society. He has been involved in many strategic government development projects, and lately the UAE national ID project as an executive steering board member and the chairman of the technical committee. His main research interests include the application of modern and sophisticated technologies in large contexts, projects management, organisational change and knowledge management.

About the contributing authors

H.E. General Ahmad N. Al-Raisi

H.E. General Ahmad N. Al-Raisi is the Director-General for Central Operations at Abu Dhabi Police GHQ, and Chairman of the Executive Board Committee at the Emirates Identity Authority. He received his degree from Otterbein, Ohio State University in the United States, and is currently doing his doctorate research in London in the field of risk and disaster management. With projects ranging from force automation, administration and security systems, and fingerprint/PKI-based smart card systems to iris recognition, H.E. Al Raisi has championed many successful innovative and complex projects on both local and federal levels. His research interests include strategic management and innovation.

Professor Jay Bal

Dr. Jay Bal is an Associate Professor at the University of Warwick in the UK. He joined the Rover Advanced Technology Centre at the University as 'IT and Organisational Strategy' Programme Manager in 1986 as a founding member of staff and set up a programme of research and consultancy for the Centre. Concurrently he helped to develop Rover IT strategy and managed a number of key IT projects. Since joining the University Dr. Bal has devised and taught courses in Information Technology, Artificial Intelligence and on Design and Manufacturing systems in the Electronics Industry to senior managers in Hong Kong, Malaysia, India, China and South Africa as well as the UK. In the last five years he has published over ten papers in international journals, and spoken at many international conferences on aspects of e-business.

Project management in reality: lessons from government IT projects

Ali M. Al-Khouri

Abstract: This article presents some practical insights and challenges encountered during the implementation of major IT projects in the government sector in Arab countries. The primary purpose of this article is to highlight and add any identified pitfalls to the existing body of knowledge from a practitioner's standpoint, as many of the articles published in this regard are published by vendors, consultants, or academics. Each item is discussed to demonstrate how it impacts on management and the overall performance of projects. They are believed to contribute significantly towards the successful management and implementation of projects, and to be valuable lessons that should be recorded in an organisation's knowledge and watch list records.

Key words: *Project management; project failure.*

1. Introduction

It is widely accepted in the literature by both academics and practitioners that information technology projects have very high failure probabilities and that between 60 to 70 per cent do actually fail. Many other researchers argue that the actual figure might be far more frightening since many organisations tend not to disclose such experiences, due to fear of criticism either by inspection or the media (Collins, 2006; Cross, 2002; Fichter, 2003).

Perhaps this may be attributed to the fact that current information technology is more complex and diverse than several years ago as it is moving out of the back office and into more mission-critical business processes, for example, customer interfaces, electronic commerce, supply chain management and so on. (Gartner Group View, 1999). Besides, many researchers have pointed out that many of today's failures are avoidable (Avison & Wood-Harper, 1990; Bentley, 2002; Berkun, 2005; Broder; 1999; Curtis, 1998; Lam, 2003; Radosevich, 1999). They argue that many projects fail because of foreseeable circumstances and that organisation's need to pay careful attention to several factors in order to reduce failure.

The findings of this article correspond with the often quoted statement in literature dealing with failure, which is related to the fact that organisations tend to treat IT projects from purely technological perspectives, and do not give much attention to other organisational issues. Almost all challenges and pitfalls reported in this article were organisational issues related to management and people. Figure 1 provides an overview of the pitfalls.

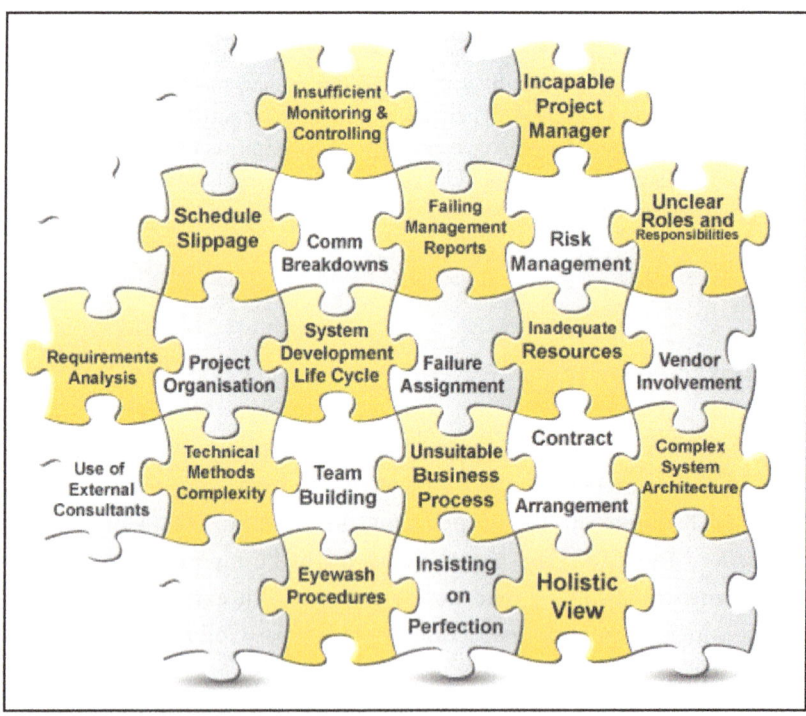

Figure 1.1   An overview of project pitfalls

The identified elements hindered the progress of projects and repeatedly delayed them from meeting planned go-live deadlines. The elements highlighted here are considered to be valuable lessons learned during the implementation of several government IT projects, and that if understood, could minimise the potential problems in management and the resulting delays in similar projects elsewhere, and smooth their implementation.

They need to be understood by the key stakeholders in projects and organisations, and not only by the project management professionals. The following sections look at each encountered project pitfall individually.

2. A holistic approach in planning projects

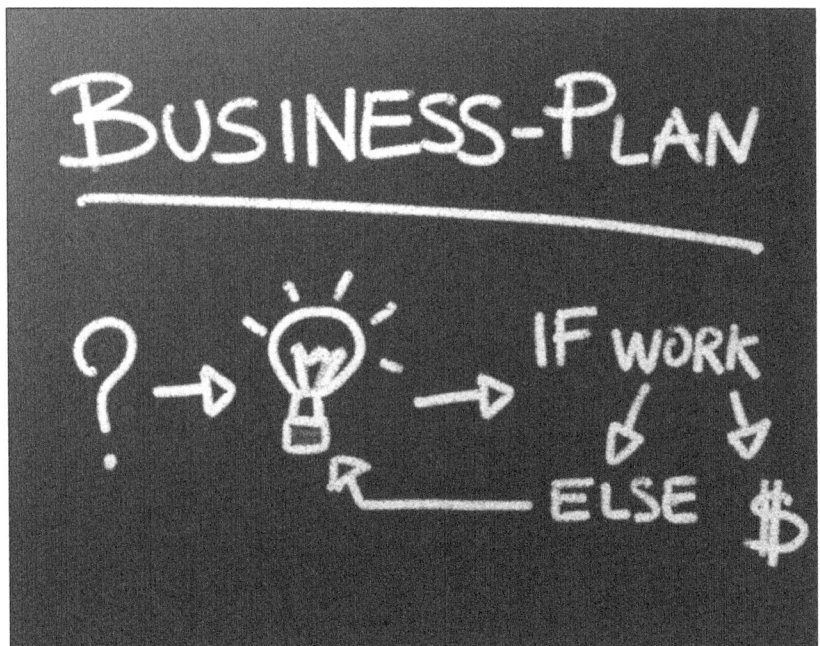

One of the key problems with IT projects especially is the fact that the basic business principles associated with each project are sometimes underestimated or totally ignored. The reason is often a total lack of knowledge, or ignorance in regard to the simple principles of sound public administration.

No project can be implemented in a smooth and timely manner if the overall business process is not determined or comprehended up front. This entails the identification in detail of the different processes, the supporting legislation, the rules and regulations to be applied, what will be required in terms of finance, human resources, and accommodation and its related requirements.

The IT divisions should only be tasked with a project's execution once these basics have been determined. For example, once the different processes required by the project have been determined in detail by business experts, it becomes easy to determine the required skills necessary to carry out all the different aspects of the project.

Personnel working on the projects should therefore consist of a blend of business as well as IT experts. Decisions regarding business issues will be made by business experts and not by IT experts. The IT experts will therefore be given a clear indication of what is required by the business experts and can apply their expertise to carry out the current project in the best possible way. Through this interaction between business and IT experts the best possible solutions and/or decisions will be implemented.

As a result of poor planning, all the relevant aspects of a project are not taken into consideration, resulting in unrealistic deadlines being set. Often when these deadlines are not reached, a scapegoat is looked for to take the blame. Thus, it is very often seen that the magnitude of a project is underestimated to such an extent that unrealistic timeframes and closing dates are set for a project. This is typically the result of inexperienced people doing the planning.

Without a clear understanding of what the project entails, it is often found that the wrong skills are appointed or they are appointed at a very late stage in the project. There is always a tendency from these appointees to re-invent the wheel. This results in endless disagreements from both sides and causes delays without real value being added to the project. It is therefore of the utmost importance to appoint the required skilled personnel on both the business and IT sides from the outset, in order to avoid these fruitless differences and delays.

Provision is sometimes not made in advance for the accommodation of equipment, for personnel working on the new project and for the public to be served in a user-friendly environment. This also leads to endless delays, and deadlines have to be continually postponed.

There should also be a clear line of communication between the project management and the sponsor of the project. Without that, vital information is sometimes 'withheld' from the sponsor, or his decisions are anticipated only to be rectified once he becomes aware of the situation.

3. The dilemma of determining realistic timeframes

In almost all of the implemented large-scale projects, each review of the projects' schedule confirmed a delay of several weeks or even months to the deadline for their formal starting dates. The following factors were the primary reasons for delays that obstructed the ability to keep to the agreed schedules:

- The amount of work that had to be performed was underestimated and the process to be followed was not clear.
- An underestimation of the requirements of the projects up front often led to an inappropriate solution offered by the vendors. This solution was then included in the contract and vendors were often, with

good reason, reluctant to deviate from the contract for fear of widening the project. The requirements should therefore be absolutely clear and signed off by a skilled business expert who should take full responsibility for the project. The proposal and its associated contract should likewise cater in detail for all these requirements. This will ensure changes to the system and the contract will be restricted to an absolute minimum.

- The identification of project activities was not easy and the time required for their accomplishment was too short.

- The approval of system specifications took longer than specified. This was mainly due to an unrealistic date set for this as a result of inadequate specifications, to the complexity of the system, and also to the unavailability of business experts, project staff and IT experts for decision-making, which consequently caused a postponement in the completion dates for many documents and project activities.

- Far too much emphasis was often placed on the security of the systems from an IT perspective, resulting in a closed system being developed by the vendors. The systems were therefore difficult to operate, not user-friendly to the public, and it was very difficult to carry out any changes at a later stage.

- Team members not skilled for the task were sometimes compelled to make important decisions on certain issues, only for these decisions to be revised. An example of this is when IT experts have to make business-related decisions and business experts have to make decisions on technical issues.

- Decisions were sometimes taken in good faith by management team members, only to be reversed when more senior members or the project sponsors become involved. It was therefore of the utmost importance to ensure that decisions on critical issues were cleared by a higher authority before they were implemented. There should be a structured procedure whereby these issues are dealt with.

- The vendor system development process was complex and could not carry out address ad hoc modifications to the system.

- Outstanding contractual issues that could not be resolved at technical levels took longer periods until they were resolved at on the executive management level.

- The legal requirements for some organisational aspects related to the sharing of data with other government organisations went through time-consuming governmental processes.

- Recruiting the required staff was a big challenge, especially for those jobs requiring highly skilled and knowledgeable candidates, and was often finalised at a very late stage.

- Communication and coordination with other government departments were a daunting prospect, and the delays in reaching a consensus and getting approval had an impact on the completion of project activities.

- Consultants working on the projects were often experts in specific areas of the IT environment but regarded themselves as universal experts even in advanced business processes. This resulted in the system often being confined to a closed and secure IT project with little room for change. This proved to be a problem as some consultants were part of the team negotiating contracts with vendors. Furthermore, western consultants had a tendency to cleverly and purposefully add additional time-consuming requirements to the system resulting in endless discussions and often requiring a revision of the whole system. This caused 'legitimate' delays and extended the project to a point where the contract of the consultant also needed to be extended for a further period of time.

Besides, project managers tended to either produce plans that were too broad in scope, with insufficient detail, or which were too detailed. Large projects had detailed schedules. However, it was found to be impractical to use detailed plans for reporting to the committee executives, who were usually interested in whether or not the project was on target, and they could not see this in the mass of detailed activities.

This requirement was sometimes difficult to obtain as project managers tried to hide potential delays from the top management. It is, however, of the utmost importance to provide the sponsors of the projects with detailed information outlining problems when they are come across in order for informed decisions to be made at the highest level. If not, these problems will be found out later with inevitable consequences.

In general, the process followed in the projects was more one of planning the project activity-by-activity. The assumption was that as soon as the sub-projects were started, more information would become available for the other activities. We call this 'management-by-luck'. So not surprisingly, the project managers were often sucked into a spiral of planning and re-planning, as they ceased to manage their projects and the plans lost their credibility.

Our close observations of the projects showed us that the project management teams did a good job in the up front planning process, but then were not able to manage the projects effectively from that point on.

This included problems with managing changes in scope, resolving issues, communicating proactively and managing project risks.

One explanation for this setback lay in the fact that even though in many projects the roles and responsibilities of the project teams were clear, their decision authority was often limited due to the level of influence and decision-making power of the project members.

There was also a tendency in the projects to focus on deadlines and perform target-led planning approaches. Too much attention was given to such dates. This affected the performance of the project members. In being concerned only about a point that lies far in the future, the project members felt that there was plenty of time to do the work. Consequently, the project activities were delayed and took longer than anticipated.

Project managers should realise, at an early stage already, if a project is underestimated, or has too many built-in security features, then inadequate provision is made by the vendor as a result, and deadlines cannot and will not be met. They should then advise the management and the sponsor of the project accordingly and also advise on a possible new strategy. By not doing so, problems are just postponed at tremendous cost and frustration.

For instance, when the steering committees or the top management were presented with the project schedule, they presumed that it would be possible to do the work more quickly and over shorter periods. One motive behind this was to please the project sponsor and to inform him that the project was on track.

To make the plan attractive, the project managers then reduced the estimates of work content and duration, and then convinced themselves and the steering committees that the new estimates could be achieved. Unfortunately this did not happen, and the plans were clear evidence proving the failure of such planning practices. This had serious implications on some of the projects.

There was also tendency to plan the projects as if the outside world did not exist. The project schedule lacked any slack or contingency timeframe. Many of the project activities were extended due to the unavailability of staff for reasons such as holidays, sick leave, training courses and seminars, and of course those skilled staff members that were appointed very late.

As is the case in any project, progress was dependent on certain decisions being made within the organisation. It was common not to give proper attention to the political factors underlying the decision coming from the top, and to underestimate the time required to study and implement such actions.

The result was that insufficient time and resources were given for many tasks. Sufficient time was not allowed for some important activities, which later impacted negatively on the schedule. Critical tasks were done inadequately and had to be redone. All these identified factors affected the planned activities, and required further re-planning. As indicated earlier, project managers were sucked into re-planning repeatedly.

4. Lack of vendor involvement in project management

Vendors often underestimated the value of participating in the project management process. They sometimes obstructed the concept of providing an onsite project office with a team to manage the accounts (contract) and the project on an ongoing basis. To a large extent, vendors were seen to play a passive role in the projects, limiting their involvement and responsibility to the implementation and delivery of the system.

Therefore, projects were primarily managed by the client's own resources and/or the consulting companies. Thus it was always a one-sided project management activity, with minimal input and cooperation from the vendors. The vendors, therefore, were constantly trying to stick to their proposed solutions in terms of the agreed contract and it was always difficult to obtain the required co-operation.

In fairness to the vendors it should be realised that changes involve development work which could be time consuming and costly, and if not limited to the essentials it may easily become an ongoing process which will lead to the vendor not making any profit.

Despite the fact that it was in the vendors' own interest to work very closely with the clients in order to focus on the same goal as a team, it was a common view that their responsibility was limited to the development of the system, and not to the management of the other project activities. This created a communication gap in the projects, as it also contributed to delays in other project activities which subsequently took longer periods to be completed.

5. Requirement analysis

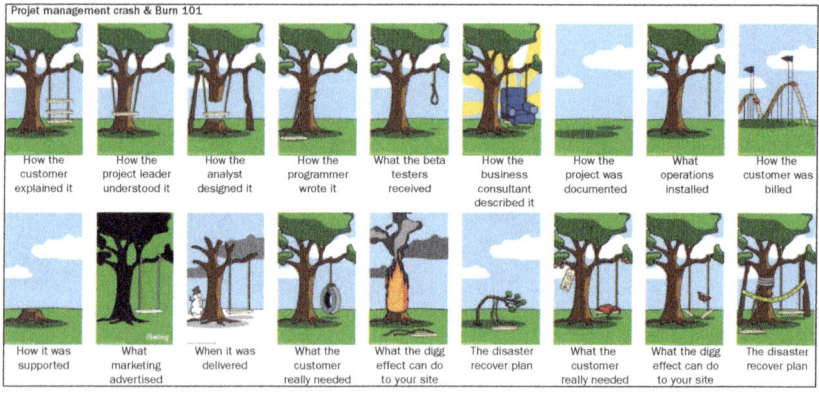

Even up to the final stages of the implementation phases of many projects, requirements were still not 100 per cent finalised. The systems first presented for testing were implemented in various versions. Technical teams normally accepted the systems because of management pressure to meet operational deadlines.

One reason for unfinished requirements up to that point was due to their not being properly outlined or clarified from the beginning. In any project there will always be cases where requirements change and/or new ones are discovered after the ideas have been conceptualised and materialised into a tangible system (see also: Avison and Fitzgerald, 2003; Checkland, 1999; Checkland and Holwell, 1998; Curtis, 1998; Mumford, 1986; Wilson, 1990), and there must always be a dynamic control process to manage such changes to the system.

It was a fact that the vendors in many projects took an overall approach at very complex systems and expected that there would be no or very little comeback in terms of changes to these systems. Instead of working closely with the clients and providing ongoing feedback and involvement in the development process (especially the user interface), it was often a one-sided effort that was undertaken by the vendors' own technical staff.

Though requested many times, some vendors did not see any point in presenting their approach to explain to the clients how they intended to meet the requirements during the pilot implementation and their overall development strategy from that point. In many other projects, the vendors' view was centred around the concept of 'tell me your requirements, and we will develop it for you'. This resulted in many heated discussions between the clients and the vendors, especially when the latter were requested to put forward business and technical solutions to certain requirements during the projects.

One of the solutions presented by some client project teams to overcome this difficulty was to have a full-time requirements specialist on site at the clients' place of work during the pilot phase and specification of the next version of the system. This solution was not welcomed by the vendors, and was seen as an added cost to the projects which they could not afford because of project delays and their current losses as a result. Perhaps the following sections will provide more explanation for the inability to clearly outline and finalise requirements.

6. System development life cycle

Any development project, however small, needs to go through (Avison and Fitzgerald, 2003; Checkland, 1999; Curtis, 1998; Wilson, 1990). This is to evaluate whether the client's requirements and expectations have been met and whether the developed system is optimal, efficient and without bugs (ibid). Iterations are not supposed to be a luxury, but rather a necessity in any development project (Avison and Fitzgerald, 2003; Crain, 1992; Curtis, 1998; Harry, 1997; Olle et al., 1991).

Some vendors seemed to like the idea of a single iteration of the system, with a user interface which was developed in isolation. Although some implemented systems went through one technical iteration, it was a common view of the technical committees that only once the public has become involved with the systems, can the final iteration be developed for the first working production version.

The importance of this should be viewed against the backdrop of the fact that it in some instances it was a completely new service that was rendered to the public for the first time. Ample provision should have been made for possible changes resulting from the public's interaction with the system.

The time it took the vendors to make the changes to the system was of concern in the main projects. There were too many channels to go through in order to have a change made to the system. Although this was healthy in a very rigid and robust environment with long time-scales, it was believed by all the project teams that a public system would not tolerate such delays.

One of the recommendations put forward was that the vendor either moves a group of the developers to the client on site in order to handle minor changes to the system with a very short turn-around time, or that the client employs some of the vendors' developers directly for a period based on a time and material principle. This again was not seen of value to the vendors and that the clients were overreacting!

Either way, the systems could not accept changes through a project or major evolution process, and there was a need to have a mechanism to implement minor changes at very short notice in the form of service or feature packs.

The technical committees, throughout the period of the projects, criticised the vendors' approach to project management and system development. For instance, from a project management point of view, many of the vendors did not make any effort to explain their project management methodology despite continuous requests from the clients.

In addition, the project deliverables were arranged and structured in a 'lot' format where the phases were clear-cut. The rigid linear approach adopted by some vendors to develop their systems, was based on the concept of executing the project phases in a sequential fashion, with output from each phase triggering the start of the next phase, and with the assumption that business requirements once 'work-shopped' and documented, were finalised.

This development process might be appropriate to some project areas, but not to the development of public systems where it not only involves certain end-users operating the system, but in fact a quite large portion of the public. This development approach was envisaged as being appropriate by the vendors because it was thought to give them the possibility of detecting requirements quickly and meeting the requirement that projects were supposed to be completed within agreed time-frames.

Although the linear approach can be useful as a framework within which the main activities and phases may be organised, the vendors needed to incorporate iteration into their system development process to address the problem of incorrect and changing requirements due to user uncertainty. The vendors' approach had to take into account that requirements usually change as projects progress and as the users come to understand the implications of the requirements.

The iterative approach was not adequate to address the fact that users do not have a fixed requirement at early stages of a project and that experimentation with a real system enables users to better define, refine, and communicate their requirements (Avison and Fitzgerald, 2003; Checkland, 1999; Crain, 1992; Curtis, 1998; Harry, 1997; Olle et al., 1991; Wilson, 1990). However, revising the development approach was observed to be associated with extra cost, which the vendors were not willing to consider as an option at all.

We need to refer back to the fact that it is widely known in the field of information technology that the rigid linear approach to systems development has been the prime factor behind the failure of many IS/IT systems because of its rigidity and the assumptions behind the arrangement of its phases. IT divisions are sometimes unfairly blamed for the failure of projects, while it is actually a failure on the part of the business divisions not clearly indicating the business requirements which should be met in association with IT projects.

7. Use of external consultants

It was realised that the steering committees and top management did not always understand the depth of the changes required. Senior executives rarely concerned themselves with the details of the projects. Therefore, they hired consulting companies and sometimes individual consultants as well to deal with these details.

This again had a great impact on the project triangle in respect to the cost of obtaining the new consultants. It was also found that the consultants caused several delays to the project since they required a great deal of time to analyse management requirements, and come up with their solutions.

Time and cost were not seen by the executive management to carry the same value as the quality of the final products, which they were more interested in. The next section provides some examples from the implementation of projects.

8. Insisting on perfection

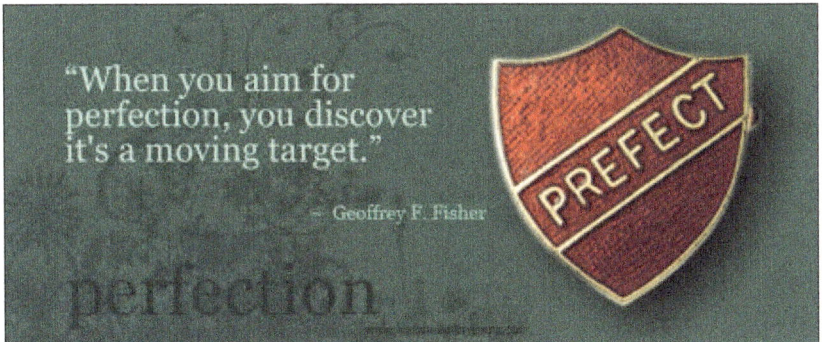

"When you aim for perfection, you discover it's a moving target."

– Geoffrey F. Fisher

There was a great tendency by some highly ranked project members to insist on identifying every conceivable risk and controlling all possible deficiencies. This led the team to study and review project specifications and other technical documentations in great detail, as some activities in large projects took almost a year to finalise in order to accommodate the different view points. From our point of view the processes and activities were straightforward and did not require much experience to set up.

There was often a tendency in the project to over-emphasise the technical aspects and ignore the organisational aspects. People find it easier to imagine concrete, technical tasks rather than abstract, organisational ones (Berkun, 2005; Burnes, 2004; Kerzner, 2004; Keuning, 1998).

Workshops on technical specifications also tended to take a long time to be finalised, where many project members were very keen to enter into discussions on technical matters and to initiate solutions before even being clear on their purpose and what they were meant to achieve. All this had a great impact on the project plans and the planned go-live dates.

In an attempt to understand and provide an explanation for such behaviour, this aspect was investigated further. It was found that the projects involved staff from very technical and operational backgrounds.

It is a fact that operationally-minded project members often tend to add more and more to the project, as they can think of hundreds of improvements that, in their minds at least, must be made (Lam, 2003; Marrison, 2002). In fact, this added more value and strengthened the project and its specifications, but it also crippled the projects and seriously impacted on the project triangle (cost, quality/specification, and time). This means that identifying each change and tracking its impact on the project triangle was a major challenge to the actual project management exercise.

Another explanation found for such behaviour was that people normally seek, unconsciously at times, to establish their identity via the project they were involved in (Marrison, 2002; O'Toole and Mikolaitis, 2002).

It was found that some members tried to make their individual mark on their projects by championing such causes as new ideas and approaches to project specifications, methodology and quality assurance programs. In one project, a member was insisting on changing a particular product specification where if carried out, the product would have lost its international certification!

Some of the consultants working on IT projects, with the knowledge and support of senior management, made the project their own, wanted to manage it and in the process tell project managers and team members which direction should be followed. This resulted in heated arguments with resultant delays and the postponing of decisions as further deliberation with higher management was often required.

9. Contract arrangement

Many disputes took place with vendors during the projects, and many of them were due to the contracts being unclear about the development methodology of the system as the contract articles were interpreted in different ways. All contracts were well written from a legal perspective. However, they lacked technical details.

The project deliverables in some contracts were organised in a 'lot' format. This was based on the traditional linear system development approach that some vendors were never willing to change or compromise as explained earlier. The projects' scale and complexity did not allow this area to be addressed thoroughly at the time of writing the contracts.

10. The complexity of technical methods

The methods and tools employed by the vendors to describe the project processes and technical aspects contained a lot of jargon which was found to be difficult to comprehend by the project teams, who in turn created a lot of confusion and hampered communication and cooperation. Project members found many business processes very ambiguous and not clear on certain delivery deadlines where sometimes even the vendors' own staff had difficulty in making them explicit.

Structured diagrams such as data flows (DFD's), if used, would have clarified the logical relationships and the flow of data and processes in the system. However, some vendors' reply was that this was not needed. The projects' members and the end users continued to struggle to find out how the system and its sub-components worked. This led the projects' teams to develop their own versions of data flow diagrams and charts based on their own interpretation of the business processes and systems.

11. Unsustainable business processes

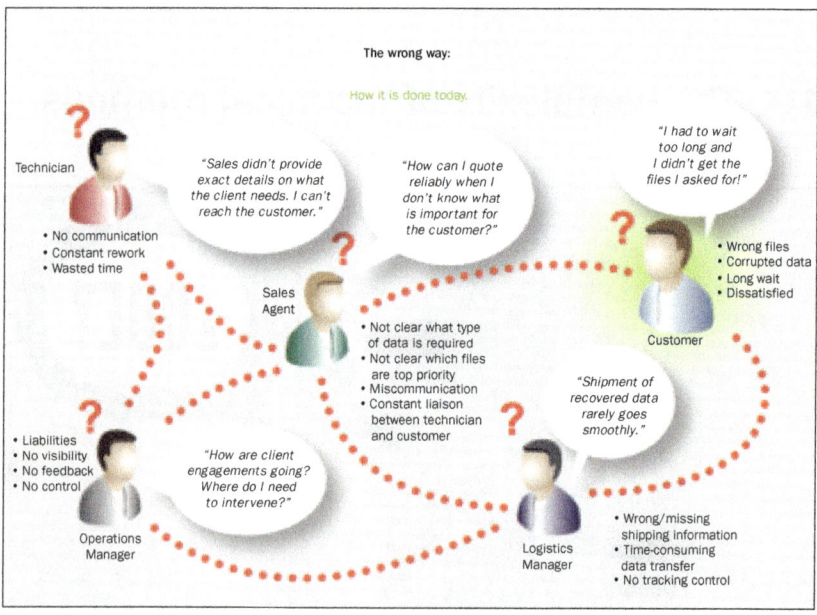

In many projects, the systems' functioning was very much determined by the consulting companies and the individual consultants who all came from western countries and underestimated many cultural aspects. It is worth mentioning, that the projects were operated more as an IT PROJECT from day one with a lot of emphasis on security and technical aspects.

An IT solution was always the approach to any business problem that came to the fore. This was why there were so many unresolved issues in all the projects from a business perspective during the testing of the system and the pilot operation, which required many manual organisational procedures to be adopted to cover up the systems' deficiencies. The vendor not being able or willing to affect changes that were identified in a timely manner further aggravated this.

12. Complex system architecture

Many of the developed systems lacked flexibility in terms of their ability to allow ad hoc programming changes to be made to their structure in order to meet requirements. The systems were developed in such a complex way that they made technical, business, and other system behavioural changes very daunting to the vendors, as they struggled to figure out how to manage and implement such changes.

In fact, adjustments to the system were always an issue and in most instances impossible to perform. The complexity challenge was due primarily to the over-use of security functions and complex programming structures, which considerably reduced system flexibility and increased intricacy.

13. Communication breakdowns

Language in the projects was clearly an issue, and many misunderstandings were the result of language constraints. The agreed language of communication both verbally and written between the project participants was English.

However, it was important to consider in some projects that the clients and the vendors had different mother tongues. The vendor experienced difficulties when participating in or facilitating workshops with staff members who were not trained to conduct such sessions (e.g., training and testing); who were not fluent in English; and who usually got emotionally involved in serious discussions, arguments, and conflicts on many occasions because of language barriers and misinterpretation of conversations.

14. Blame for failure

In some projects, the management's attention was shifted away from primary project performance factors and was focused on finding someone or some group to blame for the continual project delays. The focus was often not on error detection, prevention, or even mitigation: it was on who to blame for failure, leading to more disputes over culpability between the clients and their consulting companies.

More time was spent dealing with disputes, and less time on managing the project. Instead of focusing on the project processes and deliverables (the main task that they were brought in for), some individually hired consultants also played a key role in supporting the client to fight the consulting companies. These fights allowed some vendors in turn to have a tranquil and stress-free time throughout the projects.

15. Risk management

In projects, the risk assessment process is performed not simply to reveal potential risks, but to point out the types, locations, and strengths of the controls needed (Gilbreath, 1986). Risk management is the ability to recognise a threat, accurately assess its implications and develop approaches to mitigate the risk in a cost-effective manner (Crouhy et al., 2001; Lam, 2003; Marrison, 2002; Wideman, 1998).

A general analysis of the risks associated with the projects was performed at an early stage to identify them and weigh up their potential impact. During the initiation of the projects, the project teams were able to foresee the risks involved in a fair amount of detail. After that, however, the vision and value of the predictions concerning risk diminished.

One reason for this was that the project teams were exhausted by the daily workload and the long working hours. Their thoughts and concerns were more directed towards completing their assigned tasks. Thus, little or no time was given to perform the much needed risk assessment throughout the project.

16. Project organisation

People are, without a doubt, every project's most valuable and most perishable resource. They have to be managed, and their management is usually the biggest challenge to project success (Berkun, 2005; Binney and Williams, 1997; Burnes, 2000). Failure to clarify responsibilities and the principles of cooperation usually result in resources that are unavailable when required (Finney, 1999; Frame, 1999).

It was found that roles and responsibilities were not allocated according to individual strengths and expertise, but were based on their positions in the organisations. Not much attention was given to setting up good communication systems. There was a clear need to set up a useable collaborative environment where project members could have access to project information and communicate with each other throughout.

17. Unclear roles and responsibilities

All projects had a defined structure. However, the roles and responsibilities of all project members were not clearly defined. This led to reduced project momentum and resulted in the loss of valuable project time in discussing principles that should have been clarified at the outset.

Many of the project members wasted time doing work that was not their responsibility, especially the consultants. This pitfall would have been avoided by having a clear responsibility chart. The lack of a formally appointed 'owner of' some projects had a severe impact on the overall progress and performance, as project members ceased to work in harmony thus their efficiency decreased.

18. Team building

A motivated team in which all members are equally involved and can rely on each other is a key factor for success (Larsen, 2004). Organisations therefore need to devote time for the planning and development of a positive project culture (Harry, 1997; Ives and Olson, 1985; Newman and Sabherwal, 1996). People management was not an easy task in all the projects. Projects members and teams had different starting points and personal aims. Most of the project teams included senior managers and decision-makers at the same level.

This caused a lot of conflicts at different stages of the projects as many of them were not all working by the same rules and procedures and had their own agendas. This again weakened cooperation and reduced the potential for project members to benefit from each other. It also reduced the projects managers' flexibility as it was hard to transfer people from one activity to another.

19. Inadequate resources

For any business endeavour to succeed it must be blessed with the right amount of resources at an acceptable level of quality and skills (Hitchin and Ross, 1994). The rate of staff recruitment in all projects was far too slow to bring the necessary staff across a broad spectrum of vital positions to the level of knowledge required by the deadlines. The workload (managerial and operational) continued to be done by a group of people too small in number and not all of them were dedicated to the projects.

20. The incapable project manager

Managing complex large-scale projects requires organisational and technical skills (Huber, 2003; Mullaly, 2003; Schneider, 1999). It requires dealing effectively with new technology, new business processes, and changes in organisational structures, standards, and procedures (Harry, 1997; Ives and Olson, 1985; Newman and Sabherwal, 1996). Unless the project manager is sensitive to the impact of each of these elements on the project as a whole, he or she is likely to get caught in conflict situations (ibid).

The beginning of all the projects was a honeymoon period, when all the participants felt great about the project. It was when the project teams began to develop substance in their recommendations that difficulties arose. The project managers were supposed to be convincing and persuasive to sell the overarching goals in order to obtain the continuous support of the project members. The problem was that the project managers, often from consulting companies, were viewed in some projects as people standing next to and supporting the vendors.

Some project managers, in an attempt to hide their project management shortcomings, were always trying to involve themselves in low-level activities that could have been accomplished by the most junior staff in the projects. There were also moments when the project managers became frustrated with the project teams, and began to think of them as people who needed to be disciplined and made to pay more attention to the project, and so started reporting them to their committees.

The workshops were sometimes turned into centres of operations where the noise level was high, especially from the project managers, who should have been maintaining control. They were incapable of controlling the teams partly because of the equal power of the technical teams. This set up an increasingly negative atmosphere in the projects. It seemed they were heading towards serious trouble.

A few project managers decided to leave shortly before the first pilot operation start date. Although some senior team members took over their positions upon their leave, it was apparent that they left behind a gap in knowledge in many project areas that they accumulated during their assignment. Those projects went through a struggle for a while until the new members built up the required knowledge.

21. 'Eyewash' procedures

One common element seen in many projects is the adoption of so-called 'eyewash' procedures that do not give meaningful directions, or guidelines (Gilbreath, 1986). One may argue that as long as these procedures do not prevent performance, understanding or control of project work, they present little problem (ibid.). However, what usually occurred in the projects was just the opposite: confusion and wasted resources.

The consulting companies presented many procedures that were supposed to be followed in the projects. In reality, only a few were followed, for example, changes in requests, meeting minutes and risk sheets, to name a few. The fact that there were too many procedures and templates confused many project members and led them to develop their own ways of doing things. This again created more confusion, especially since some projects' documentation followed different standards.

22. Failing management reports

Information plays a dual role in any project, that of a valuable resource and an essential tool. Management reports are only as good as the information they contain and that information has value only when it facilitates analysis (Garreth et al., 2000; Gilbreath, 1986; Laudon and Laudon, 1998; Mullins, 1996; Wysocki, 2000). It does not matter how they are structured, how frequently they are produced, or how many ways they can slice and graphically depict the information pie, reports that do not facilitate analysis do not deserve the management's attention (ibid).

For long periods, the projects' monthly reports did not carry the desired quality of information to enable the committees to maintain an overview of the projects' progress, or to understand the potential risks the projects may have been heading towards. In fact, the submitted reports were lengthy and contained lots of data that led the committees to abandon reviewing the monthly reports altogether.

23. Conclusion

Project management is a research topic on which there is much literature. However, recent studies show that the failure rate in the IT industry is still a continuing story. No overarching framework or methodology to guide projects to success has yet emerged. This paper has presented some lessons from many government IT projects where the challenges reported herein obstructed the maintenance of the triangle of the projects' schedules, budgets and qualities equally.

The reported pitfalls were almost all organisational issues related to management and people. Thus, these factors need to be comprehended by both project management professionals and key executives/owners in organisations.

References

1. Avison, D. E. and Wood-Harper, A.T. (1990) 'MultiView – an exploration in information systems development'. USA: McGraw Hill.
2. Avison, D.E., Fitzgerald, G. (2003) 'Information Systems Development: Methodologies, Techniques and Tools' (3rd Edition). McGraw-Hill: London.
3. Bentley, C. (2002) 'Practical PRINCE2'. USA: The Stationery Office Books.
4. Berkun, S. (2005) 'The Art of Project Management'. USA: O'Reilly.
5. Binney, G. and Williams, C. (1997) 'Leading into The Future: Changing The Way People Change Organisations'. London: Nicholas Brealey Publishing.
6. Broder, J.F. (1999) 'Risk Analysis and the Security Survey' (2nd Edition). Boston: Butterworth-Heinemann.
7. Burnes, B. (2004) 'Managing Change' (4th edition). London: FT Prentice Hall.
8. Checkland, P. (1999) 'Systems Thinking, Systems Practice'. Chichester: John Wiley and Sons.
9. Checkland, P. and Holwell, S. (1998) 'Information, Systems, and Information System: Making sense of the field'. Chichester: John Wiley and Sons.
10. Collins, T. (2006) 'Government IT: What happened to our £25bn?' *IT Management, Politics & Law*, <*www.ComputerWeekly.com*> Available at: *http://www.computerweekly.com/Articles/2006/10/30/219476/governmen t-it-what-happened-to-our-25bn.html*
11. Crain, W. (1992) 'Theories of Development: Concepts and applications' (3rd Edition). New Jersey: Prentice Hall International.
12. Cross, M. (2002) 'Why government IT projects go wrong'. Available at: *http://www.itweek.co.uk/computing/features/-2072199/whygovernment-proje-cts-wrong*
13. Crouhy, M., Mark, R. and Galai, D. (2001) Risk Management. USA: McGraw-Hill.

14. Curtis, G. (1998) 'Business Information Systems: Analysis, Design, and Practice' (3rd Edition). USA: Addison-Wesley.

15. Fichter, D. (2003) 'Why Web projects fail'. 27 (4): 43–45.

16. Finney, R. (1999) 'Monitoring an Active Project Plan'. *Information Technology Management WEB*. Available from: *http://www.itm-web.com/essay012.html*

17. Frame, J.D. (1999) 'Project Management Competence'. Jossey Bass: San Francisco.

18. Gareth, R., George, J.M.; Hill, C.W.L. (2000) 'Contemporary Management' (2nd Edition). USA: McGraw-Hill.

19. Gartner Group View (1999) 'Are all IT projects doomed for failure'. *Asia Computer Weekly*, September: 1.

20. Gilbreath, R.D. (1986) 'Winning at Project Management: what works, what fails and why'. New York: John Wiley and Sons.

21. Harry, M. (1997) 'Information Systems in Business' (2nd Edition). Great Britain: British Library Cataloguing in Publication Data.

22. Hitchin, D. and Ross, W. (1994) 'Achieving Strategic Objectives: The Role of Human Resources and Organisational Development'. USA: Addison Wesley Longman Publishing Co.

23. Huber, N. (2003) 'Hitting targets? The state of UK IT project management'. *Computer Weekly*, November.

24. Ives, B. and Olson, M. H. (1985) 'User involvement and MIS Success,' *Management Science*, 30 (5): 586–603.

25. Kerzner, H. (2004) 'Advanced Project Management: Best Practices on Implementation'. NJ: Wiley and Sons, Inc.

26. Keuning, D. (1998) 'Management: A Contemporary Approach'. London: Pitman Publishing.

27. Lam, L. (2003) 'Enterprise Risk Management: From Incentives to Controls'. NJ: Wiley and Sons, Inc.

28. Larsen, E.R. (2004) 'People: The Key to Successful Project Management'. *Chemical Engineering Progress*, 100 (9): 55–58.

29. Laudon, K. and Laudon, J. (1998) 'Management Information Systems – New Approaches to Organisations & Technology'. New Jersey: Prentice Hall, 334–379 and 624-653.

30. Marrison, C. (2002) 'The Fundamentals of Risk Measurement'. USA: McGraw- Hill.

31. Mullaly, M.E. (2003) 'The Accidental Project Manager: Coming In From the Cold', *<www.gantthead.com>* Available at: *http://www.gantt-head.com/article.cfm?ID=165059*

32. Mumford, E. (1986) 'Participation in Systems Design: What it can Offer'. paper presented at SERC/CREST Advanced Course, Loughborough University.

33. Newman, M. and Sabherwal, R. (1996) 'Determinants of Commitment to information Systems development: a longitudinal investigation'. *MIS Quarterly* 20 (1).

34. Olle, T.W., Hagelstein, J., Macdonald, I.G., Rolland, C., Sol, H.G., Van Assche, F.J.M. and Verrijn-Stuart, A.A. (1991) 'Information Systems

Methodologies: A Framework for Understanding (2nd Edition). Wokingham: Addison-Wesley.

35. O'Toole, W. and Mikolaitis, P. (2002) 'Corporate Event Project Management'. NJ: Wiley and Sons Inc.

36. Radosevich, L. (1999) 'Measuring Up: the importance of metrics in IT project management'. *CIO Magazine*, September. CIO Communications, Inc.

37. Schneider, P. (1999) 'Wanted: ERPeople Skills'. *CIO*, March: 30–37.

38. Wideman, R.M. (1998) 'Project and Program Risk Management: A Guide to Managing Project Risks and Opportunities'. USA: Project Management Institute.

39. Wilson, B. (1990) 'Systems: Concepts, Methodologies and Applications' (2nd Edition). Chichester: John Wiley and Sons.

40. Wysocki, R. K., Beck, R. and Crane, D.B. (2000) 'Effective Project Management'. New York: John Wiley.

An innovative project management methodology

Ali M. Al-Khouri

Abstract: This paper presents a project management methodology - developed part of an engineering doctorate research at Warwick University – for managing strategic and large scale IT projects. The methodology was mainly tested in 4 countries. The research demonstrated that by following a formal structured methodology, governments will have better visibility and control over such programmes. The implementation revealed that the phases and processes of the proposed methodology supported the overall management, planning, control over the project activities, promoted effective communication, improved scope and risk management, and ensured quality deliverables.

Key words: *Project management, project methodology*

1. Introduction

Recent studies estimated the cost associated with implementing large-scale government IT projects to scale up to multi-billion US dollar enterprises. (Fontana, 2003). Obviously, the nature, size and complexity of these projects raise the probability of failure. This is in reference to the accepted phenomenon in literature by both academics and practitioners that information technology projects have a very high chance of failure, and that between 60 to 70 per cent do actually fail. Many other researchers argue that the actual figure might be far more frightening since many organisations tend not to disclose such experiences due to fear of criticism by audit or the media (Dorsey, 2004; Fichter, 2003).

By and large, the knowledge required to succeed with IT is complex and rapidly changing. It is noted that the examples in existing literature are rarely of the size and complexity of those executed in the government sector. Proceeding without understanding or managing the inherent risk in such projects will obviously lead to higher probabilities of failure.

This paper presents an overview of a project management methodology that was implemented in four countries to support the management and control of the project phases. This paper is structured as follows. First some recent studies on IT project failure are highlighted along with the factors leading to such results. Then the field of project management is briefly explored to pinpoint the need for a methodological approach to managing large IT projects. The process followed, underlying principles, and an overview of the proposed methodology phases are provided next. A synopsis of the implementation of methodology and its value are outlined in the final two sections.

2. IT project failure

In line with the above statistics, it is estimated that between 20 to 30 per cent of industrialised national government IT projects fall into the total failure category; 30 to 60 per cent fall into the partial failure category; and that only a minority fall into the success category (Heeks, 2003). Studies indicate that large-scale projects fail three to five times more often than small ones (Charette, 1995).

Such failure can impede economic growth and quality of life and the cost of failure may become catastrophically excessive as societies come to rely on IT systems that are ever larger, more integrated, and more expensive (ibid). Many researchers pointed out that a lot of today's failures are avoidable, that many projects fail because of foreseeable circumstances and that organisations need to give careful attention to several factors to avoid failure (Avison and Wood-Harper, 1990; Bentley, 2002; Berkun, 2005; Broder; 1999; Curtis, 1998; Lam, 2003; Radosevich, 1999).

Among the widely quoted factors contributing to failure is the fact that organisations tend to treat IT projects from purely technological perspectives, and do not give much attention to other organisational and

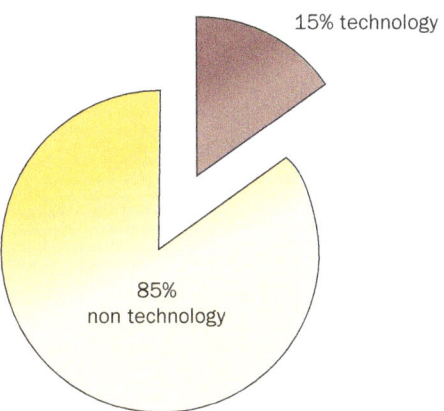

15% technology

85%
non technology

Figure 2.1 Critical success factors - technology v. non-technology

management issues. Some literature shows that technology can contribute as little as 15 per cent to the overall success of projects (see also Figure 2.1), where the remaining 85 per cent is dependent on wider organisational issues related to people, data, and management.

Research points to the fact that one of the principal causes of information system failure is when the designed system fails to meet the business requirements or improve the organisational performance. Figure 2.2 below illustrates an example of how a user's requirements might be interpreted, not only at the requirements analysis stage but throughout the project.

3. Project management

The argument of this research mainly advocates that by following a disciplined methodology that sets standards for all phases of a project there is more likelihood of increasing the chances of success. Project management is viewed as the art of defining the overall management and control processes for a project (Devaux, 1999; Garton and McCulloch, 2005; Stankard, 2002).

Project management works with differing degrees of success in different industries, different organisations, and on different projects. What is undeniable is that industries have been much more successful when project management is used than when it was ignored (Devaux, 1999; Ireland, 1991).

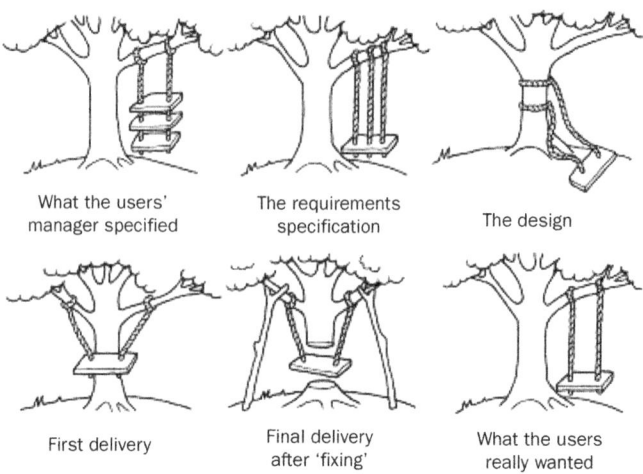

What the users'
manager specified

The requirements
specification

The design

First delivery

Final delivery
after 'fixing'

What the users
really wanted

Figure 2.2 Interpretation of user requirements

A project management methodology that takes into account the factors of success and failure in the field of IT projects is more likely to increase the probabilities of success for the project (see for example: Avison and Fitzgerald, 1998; Curtis, 1998; Flynn, 1998).

Looking at the existing available methodologies, the literature perceives the field as a jungle with a large and confusing variety of approaches in existence (Avison and Fitzgerald, 1998). It is estimated that over a thousand brand name methodologies exist worldwide (Jayaranta, 1994). Charvat (2003) found in an analysis of 18 different methodologies that:

(1) Some focus purely on the technology itself;

(2) others focus more on a generic project management approach.

Such researchers argue that organisations need to carefully assess the methodology based on the organisational requirements and that it is the project size and complexity which necessitates the use of a fitting methodology (Berkun, 2005; Charvat, 2003; Radosevich, 1999; Verrijn-Stuart, 1991; Gilbreath, 1986). From a practical point of view, there is no one methodology that guarantees success, but by employing one, an organisation will have a structured set of concepts to handle each step – from understanding the business requirements to the development of the system – of the project (Avison and Fitzgerald, 1998; Crain, 1992; Curtis, 1998; Harry, 1997; Ives and Olson, 1985; Newman and Sabherwal, 1996; Olle et al., 1991).

The following section describes the process followed in the development of the methodology that was later implemented to manage the UAE programme.

4. Crafting the methodology

Researchers have continuously emphasised the need for organisations to seriously analyse failed or out-of-control IT projects and the associated challenges. Nonetheless, research to date has found no single explanation for system success or failure. Nor does it suggest a single or a magic formula for success.

However, it has found different elements leading to project success or failure. These elements were more or less presented in the Standish Group *CHAOS 2001* report as shown in Table 2.1. Comparing these factors from the Standish report to the literature, the most common factors that contributed to project success or failure were:

- Management commitment
- Business strategy focus

Table 2.1   Indicators found between successful and failed projects

Successful projects	■ User involvement ■ Executive management support ■ Clear statement of requirements ■ Proper planning ■ Realistic expectations
Challenged projects	■ Lack of user input ■ Incomplete requirements and specifications ■ Changing requirements and specifications ■ Lack of executive support ■ Technical incompetence
Failed projects	■ Incomplete requirements ■ Lack of user involvement ■ Lack of resources ■ Unrealistic expectations ■ Lack of executive support ■ Changing requirements and specifications ■ Lack of planning ■ Did not need it any longer ■ Lack of IT management ■ Technical illiteracy

- Requirements definition
- Complexity management
- Changing targets
- Formal methodology
- Project management
- Planning
- User involvement
- Risk management

These factors were taken into consideration when designing the methodology. In its development and implementation the underlying principles were based on theories and practices coming from two subject areas:

(1) Project management, and

(2) system development.

In addition to the above elements, the methodology has been developed to address the core needs identified for supporting and improving the following:

1. Concept development
2. Overall project portfolio management
3. Management of stakeholders' expectations
4. Analysis of requirements
5. Quality of output
6. Utilisation of resources
7. Communication and management reporting
8. Project control and risk management
9. Knowledge management

A two-staged project management methodology consisting of nine phases was developed depicted in Figures 2.3 and 2.4. The methodology is composed of the following inter-linked processes:

1. Initiating processes
2. Planning processes
3. Executing processes
4. Controlling processes
5. Closing processes

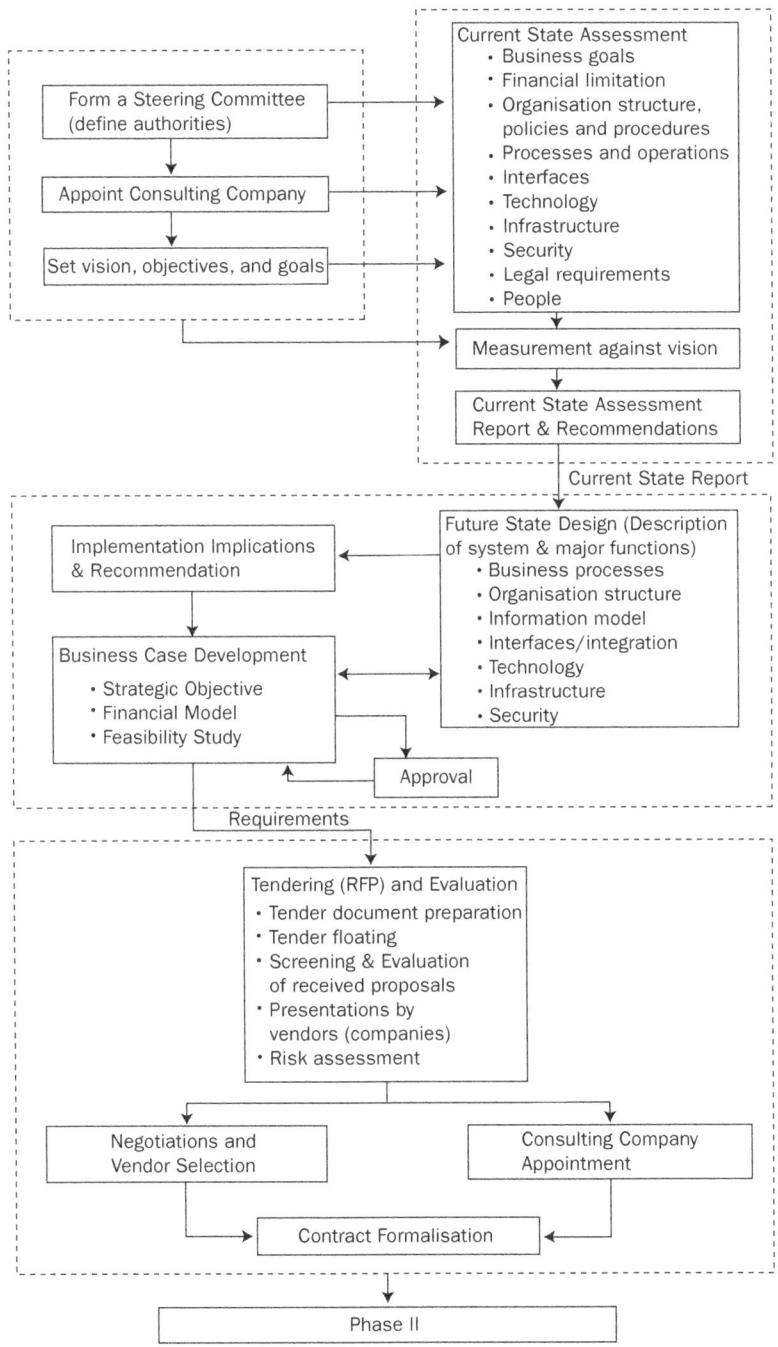

Figure 2.3 Phase one of the proposed methodology

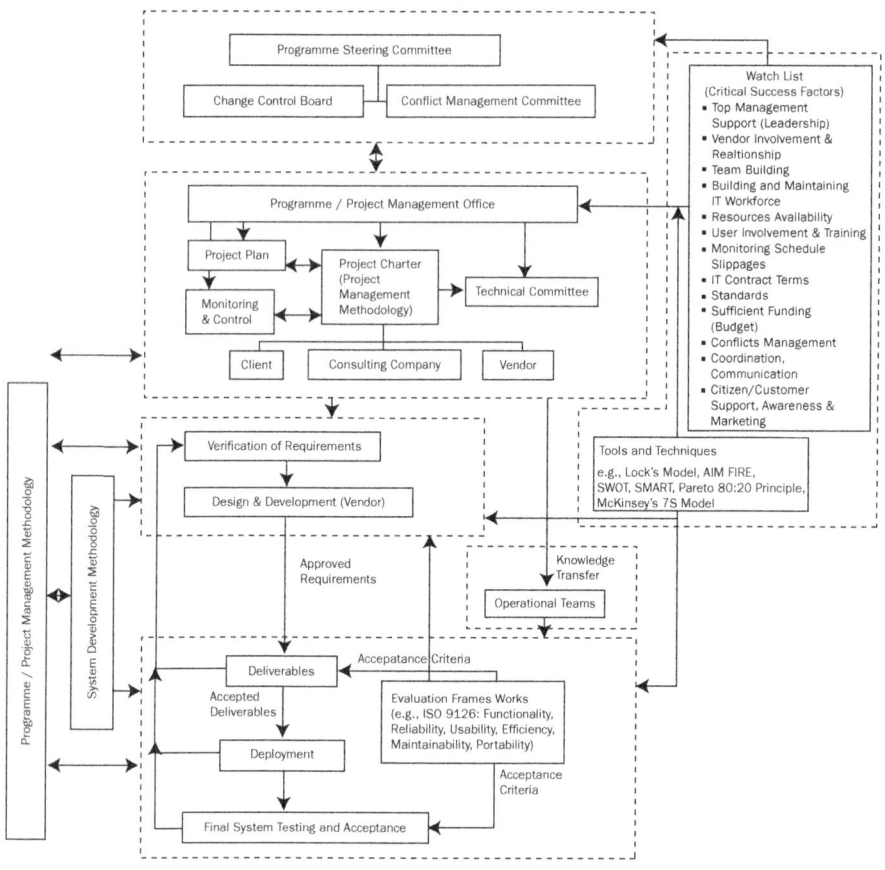

Figure 2.4 Phase two of the proposed methodology

5. Applying the methdology

The methodology was mainly tested in four large government projects. Presentations at conferences and official delegation visits to many countries worldwide contributed to the overall enhancement and supported the value of the methodology. The model was refined at several stages to address common problems identified during the implementation, and from the feedback from government officials and experts in the field.

The implementation of the methodology has revealed the following contributions to the overall project management practices:

1. Agreed and articulated project goals and objectives.
2. Staged and controlled phases with sign-offs.
3. Regular reviews of progress against plan, scope, quality.
4. Supports project and management status reporting.
5. Global overview of the project processes, beginning and end of the project, and all the work in the middle.
6. Strong management control through clear change control and conflict management procedures.
7. Promoted the involvement of management and stakeholders at different stages of the project.
8. Clear focus on defining system requirements.
9. Capturing and sharing of lessons learned.
10. Improved project control – evaluate and measure performance based on the defined scope, schedule, budget, quality of deliverables.
11. Transparent project management practices.
12. Risk management.
13. Handled project complexity.
14. Open communication channels among the project stakeholders.

Twelve areas for consideration that governments need to heed during the course of the scheme implementation were also identified:

- Having clear goals
- Appointing a project director
- Setting up a project management office
- Assessing project mangers skills (leadership skills, communication, negotiation, delegation, problems solving, change management and so on).
- Developing stakeholders commitment
- Dealing with project team anxiety
- Team assembly
- Responsibility assignment (RA) through RA Matrix
- Monitoring, evaluation and control
- Performance observation

- Planning and communicating
- Skilled team members
- Flexibility of the project plan

6. Significance and value of the methodology

The implementation results of the systems, and the feedback from the government officials and experts in the field indicated the value and significance of the methodology. The methodology was noted to incorporate flexible and easy to understand principles of project management in order to improve the planning and control of the government schemes.

There was also a common view that the phases and processes of the methodology improved the overall visibility and control of the project activities, promoted effective communication, supported scope and risk management, and ensured quality delivery. One could still argue that multiple data from multiple case studies would have provided a better indication of the reliability and value of the development methodology. However, government policies, especially in this kind of project do not allow close involvement or data dissemination. In fact, government projects tend to be hidden from the public domain in many countries.

The nature of strategic and large-scale IT programmes carries a much higher confidentiality rating. Nonetheless, to prove real usefulness and contribution as a standard for managing such programmes, it would require a significant period of time to gather objective quantitative data from different experiments. Nevertheless, the methodology has an advantage over the existing standard methodologies by the fact that it has been applied and customised to four international projects. The appreciation and feedback from these four countries, government officials worldwide, and experts in the field also demonstrates the usefulness of the methodology.

7. Conclusion

Following best practice or a particular project management methodology or framework, cannot make projects failure proof. To succeed, a successful project needs much more than a cookbook approach, especially when implementing large-scale projects. There are many issues that require management attention, and a comprehension of their possible impact is considered essential to increase the chances of a successful endeavour.

This research attempted to add value to the limited knowledge currently available to practitioners and researchers on project management methodologies, and introduced a new test project management methodology in four large-scale government IT programmes. Project management was found to be crucial in large and complex projects where attention was required to analyse and carefully respond to the implications of the slightest change.

It is important for management to realise that every project is unique, and that by repeating old experience or practices from similar past projects it will not help to accommodate the ever-changing landscape of today's projects. Successful implementation requires clear business processes, the ability to manage the system's flexibility, and the ability to cope with high complexity levels (Frame, 1999; Garton and McCulouch, 2005; Gilbreath, 1986).

Furthermore, successful implementation requires superior skill in a variety of generic business, communication, and organisational skills, in addition to knowledge of the technology being used (Haller, 1998; Ireland, 1991; Kerzner, 2004; Page, 2002). Under these circumstances, project management is argued to be essential to achieving better visibility and control over projects.

References

1. Avison, D. E. and Wood-Harper, A.T. (1990) 'MultiView – an exploration in information systems development'. USA: McGraw Hill.
2. Bentley, C. (2002) 'Practical PRINCE2'. USA: The Stationery Office Books.
3. Berkun, S. (2005) 'The Art of Project Management'. USA: O'Reilly.
4. Broder, J.F. (1999) 'Risk Analysis and the Security Survey' (2nd Edition). Boston: Butterworth-Heinemann.

5. Charette, R.N. (1995) 'Why Software Fails, IEEE Spectrum'. Available at: *http://www.s-pectrum.ieee.org/sep05/1685*

6. Charvat, J. (2003) 'Project Management Methodologies'. NJ: John Wiley & Sons Inc.

7. Crain, W. (1992) 'Theories of Development: Concepts and applications' (3rd edition). New Jersey: Prentice Hall International.

8. Curtis, G. (1998) 'Business Information Systems: Analysis, Design, and Practice' (3rd Edition). USA: Addison-Wesley.

9. Devaux, S.A. (1999) 'Total Project Control'. New York: John Wiley and Sons.

10. Dorsey, P. (2000) 'Top 10 reasons why systems projects fail,' Available at: *http://www. dulcian.com/papers*

11. Fichter, D. (2003) 'Why Web projects fail', 27 (4): 43-45.

12. Fontana, J. (2003) 'A national Identity Card for Canada' Canada, House of Commons. Available at: *http://www.parl.gc.ca*

13. Frame, J.D. (1999) 'Project Management Competence'. San Francisco: Jossey Bass.

14. Garton, C. and McCulloch, E. (2005) 'Fundamentals of Technology Project Management'. USA: McPress.

15. Garton, C. and McCulloch, E. (2005) 'Fundamentals of Technology Project Management'. USA: Mc Press.

16. Gilbreath, R.D. (1986) 'Winning at Project Management: what works, what fails and why'. New York: John Wiley and Sons.

17. Haller, R. (1998) 'Managing Change'. London: Dorling Kindersley.

18. Harry, M. (1997) 'Information Systems in Business' (2nd edition). Great Britain: British Library Catalogue in Publication Data.

19. Heeks, R.B. (2003) 'Most eGovernment-for-Development Projects Fail: How Can Risks be Reduced?' *IDPM i-Government Working Paper* 14 (Online). UK: University of Manchester. Available at: *http://www.ed-evexchange.org/eGov/faileva-l.html*

20. Ireland, L.R. (1991) 'Quality Management for Projects and Programs'. USA: Project Management Institute.

21. Ireland, L.R. (1991) 'Quality Management for Projects and Programs'. USA: Project Management Institute.

22. Ives, B. and Olson, M.H. (1985) 'User involvement and MIS Success', *Management Science* 30 (5): 586–603.

23. Jayaratna, N. (1994). 'Understanding and Evaluating Methodologies: NIMSAD A Systemic Framework'. London, MacGraw-Hill.

24. Kerzner, H. (2004) 'Advanced Project Management: Best Practices on Implementation'. NJ: Wiley and Sons Inc.

25. Lam, L. (2003) 'Enterprise Risk Management: From Incentives to Controls'. NJ: Wiley and Sons Inc.

26. Newman, M. and Sabherwal, R. (1996) 'Determinants of Commitment to information Systems development: a longitudinal investigation', *MIS Quarterly* 20 (1).

27. Olle, T.W., Hagelstein, J., Macdonald, I.G., Rolland, C., Sol, H.G., Van Assche, F.J. M. and Verrijn-Stuart, A.A. (1991) 'Information Systems Methodologies: A Framework for Understanding'. (2nd Edition). Wokingham: Addison-Wesley.

28. Page, S.B. (2002) 'Project Management Methodology: The Key To Becoming a Successful Project Manager', <*www.gantthead.com*> Available at: *http://www.gantthead.com/ar-ticle.cfm?ID=135300*

29. Radosevich, L. (1999) 'Project Management Measuring UP'. *CIO* [Online]. Available at: *http://www.cio.com/archive/091599_project.html*

30. Stankard, M.F. (2002) 'Management Systems and Organisational Performance: The Search for Excellence Beyond ISO9000'. USA: Quorum Books, Paper 2.

3

The UAE national ID program: a case study

Ali M. Al-Khouri

Abstract: This article provides some insight into the implementation of the national ID program in the United Arab Emirates (UAE). The fundamental aim is to contribute to the existing body of knowledge in the field, as it sheds light on some of the lessons learned from the program that is believed to widen the knowledge of those involved in such initiatives

Key words: *National ID; identity management.*

1. Introduction

Many countries around the world have either already implemented, or are in the process of embarking on national ID projects. The key motives behind such initiatives is to improve the identification and authentication mechanisms in order to reduce crime, combat terrorism, eliminate identity theft, control immigration, stop benefit fraud, and provide better services to both citizens and legal immigrants (see for example: (1, 2, 3, 4).

In view of the fact that these projects are unique undertakings and involve a degree of uncertainty and risk, national ID programs are perceived to carry a high level risk and so more knowledge needs to be acquired to understand the complexity of these types of endeavours. In this regard, this article aims to present a case study of the implementation of an ID card program in the UAE, and to highlight some of the lessons learned which, if considered, are most likely to support the planning and execution of similar initiatives.

This article is structured as follows. First, some primary information about the program, its strategic goals and objectives are presented. The technologies employed and the enrolment process are explained next. A short overview of the enrolment strategy is provided, and then the lessons learned are presented as a conclusion.

2. The UAE ID card program

As a result of the rapid growth of the economy and the population over the past few years in the United Arab Emirates (UAE), the government has expressed a strong determination to enhance the performance of public departments and increase efficiency in a bid to improve the co-ordination of citizens' access to public services.

The project which was kicked off in June 2003 aimed to develop a modern identity management system with two strategic objectives addressing security and economic requirements (see also Figures 3.1 and 3.2). The security objective revolves around the necessity of the government to have an integrated population register that will become the central reference point for the whole government for the purpose of population identification and service delivery.

By employing the necessary technologies, the project has developed a trusted and robust identity verification infrastructure to enhance homeland security and help the government in protecting individuals against the ever-increasing crime of identity theft.

Figure 3.1 Project strategic goals

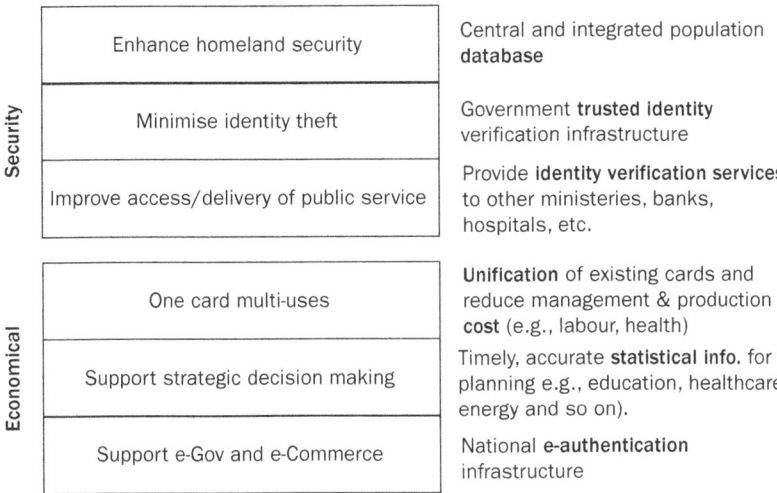

Figure 3.2 Security and economic objectives

The second strategic goal of the project revolves around supporting the government's economy. One of the key requirements is to support the country's digital economy by building a national e-authentication infrastructure which should become the basis and the backbone for e-government services and e-commerce initiatives.

Besides, having a centralised and integrated population register will assist in planning and utilising resources as it should provide timely, accurate, and statistical information for strategic decision-making and long-term planning with respect to education, healthcare, town planning, transport requirements, energy, and so on.

Another side of this objective aims to unify the existing cards in the country such as driving license, labour card, health card, and other 'entitlement'* cards. This will also have a profound impact on the economies of scale in the management and production costs of such cards.

---

* There are two main types of identity documents used in many countries. The first are often referred to as 'foundation' documents, and they include birth certificates and immigration records. These primary documents are used to obtain documents of the second type, 'entitlement' documents, such as passports, drivers' licences and other government issued documents.

3. Key operations of the national ID system

The national ID system incorporates the latest technological advances. It is designed based on a four-tier web-based architecture and complies with the latest industry standards such as ICAO for card design, x.509 for PKI, and ISO 17799 for IS security policy. The system guarantees secure communication throughout the system's national network structure by using Virtual Private Network (VPN) technology and an associated technical Public Key Infrastructure (PKI).

The fingerprint-based biometry provides the means to ensure a single identity for each applicant and to authenticate the identity of the ID card bearer. In principle, the national ID system is designed to provide three primary operations as depicted in Figure 3.3.

3.1 Population Register and Document Imaging (PRDI) management

The national ID system maintains the Population Register that records information about every UAE citizen and legal resident registered on the system and assigns a unique Identification Number (IDN) to each person. The system is currently sized to manage five million records.

It provides the means to record events such as births, marriages, divorces and deaths, as well as the updating of variable (constantly

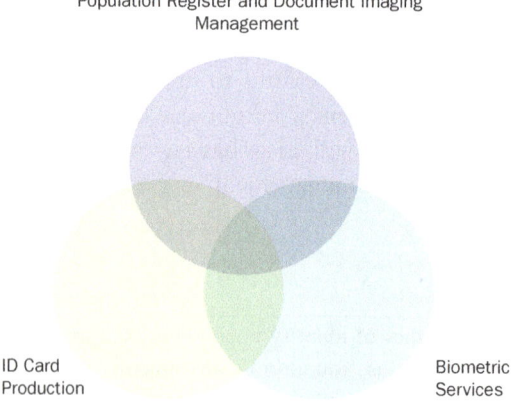

Population Register and Document Imaging
Management

ID Card
Production

Biometric
Services

Figure 3.3   **Primary operations of the national ID system**

changing) information such as address, education, employer, and so on. The national ID system also stores images of the official support documents presented during the application for an ID card or on events declaration on the Population Register.

3.2 ID card production

The national ID system includes a process for the enrolment, processing, production and delivery of ID cards. This process can be adapted for the first application for an ID card, the renewal of an expired ID card, or the replacement of a damaged, lost or stolen ID card. The ID card produced by the national ID system includes biometric fingerprint-based authentication capabilities and uses a Public Key Infrastructure (PKI) that is adapted for future e-government and e-commerce usage.

3.3 Biometric services

The national ID system provides a complete range of biometric functions using fingerprints and palm prints. The system encompasses an Automatic Fingerprint and Palm Print Identification System that provides person identification, authentication, crime solving and crime linking services. These services are used to guarantee the applicant's identity using the ID card and to ensure that a person is issued with only one ID card. The full set of biometric services is also used for law enforcement purposes.

4. The ID card

The ID card is identical for both UAE citizens and residents in terms of card design and displayed data. The card is valid for five years for citizens, and is linked to the residency permit validity for residents. The ID card includes a digital certificate with PKI capabilities. This feature constitutes one of the bases for future online identification, authentication and transactions to support e-government and e-commerce.

As depicted in Figure 3.4, it contains many security features which will make it very difficult, if not impossible, to forge the card and will allow in most cases a trained card reader to identify a fake ID card even on face value.

United Arab Emirates
Identity Card

دولة الإمارات العربية المتحدة
بطاقة هوية

Chip data
digitally
signed and
encrypted

ID Number / رقم الهوية
784-1968-7064361-5

الاسم: أحمد محمد بن عامر
Name Ahmad Mohamad Bin Amer

الجنسية: الإمارات العربية المتحدة
Nationality United Arab Emirates

60005/1

Sex M   الجنس: ذكر

Date of Birth / تاريخ الولادة
27/07/1968

Card Number   رقم البطاقة
520727521

الصلاحية / Expiry
05/03/2009

021000C8D369ED4F

Signature / التوقيع

IDARE5207275215784196870643515
6807278M0903053ARE<<<<<<<<<<<7
BIN<AMER<<AHMAD<MOHAMAD<<<<<<<

Overt:

· Guilloches and rainbow
  Ghost image
· Holographic overlay
· Variable fine lines
· Special raster
· UV ink
· Rainbow printing

Covert:

· Micro text
· Hidden text
· UV

Figure 3.4 UAE ID security features

The digital security on the ID card chip in terms of the signature and encryption features is accorded to the highest international security levels. In terms of the IT security in the contact chip, the UAE ID card follows the highest electronic standards, based on the use of asymmetric encryption and digital signatures.

5. Implementation stages and the enrolment strategy

The system was implemented in three phases (see also Figure 3.5):

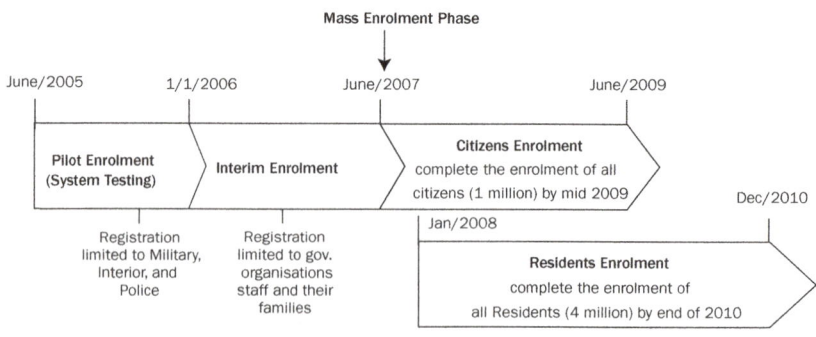

Figure 3.5 Enrolment plan

5.1 Pilot enrolment

The first live enrolment began in June 2005 as a pilot in an attempt to test the system and its capabilities. This was considered to be a good starting point since many technical and business process related issues were identified that needed this kind of system. The enrolment was stopped for more than two months to upgrade the software of the overall system to reflect the new identified requirements.

5.2 Interim enrolment

Although the system was ready for the mass enrolment phase towards the end of 2005, enrolment was limited to the registration of staff of government organisations where registration centres were available. This was due to the incompleteness of the construction projects of the majority of the registration centres across the country.

5.3 Mass enrolment phase

The registration in this phase was linked to the obtainment of certain services from the government sector such as the renewal of vehicles and passports for citizens, and the renewal and issue of residency permits for residents. With a three year enrolment strategy, the project aimed to register the whole population of the country by the end of 2010.

The registration process of new births and new residents is planned to take place directly through the Ministry of Health and Ministry of the Interior, who will interface with the national ID database, which in turn will initiate automatic requests such as first ID card applications, card issue, ID card renewal and replacement, and registered events and declarations such as marriage, divorce, birth, and so on.

6. Enrolment process

Taking into consideration the criticality and importance of the enrolment process as the new national ID card became the source document to prove an individual's identity, a robust registration process was put in place to ensure comprehensive identity verification prior to the issue of the ID card (see also Figure 3.6).

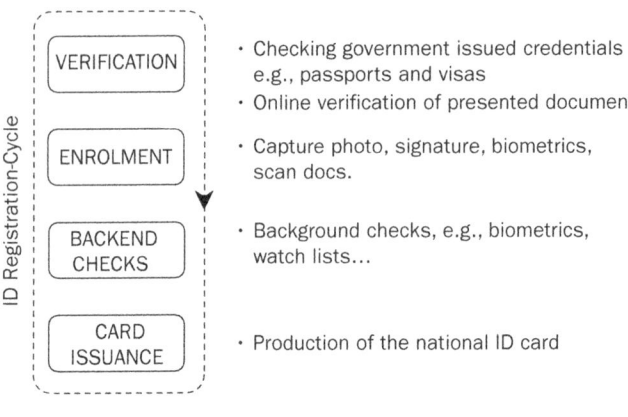

ID Registration-Cycle

VERIFICATION
- Checking government issued credentials e.g., passports and visas
- Online verification of presented documents

ENROLMENT
- Capture photo, signature, biometrics, scan docs.

BACKEND CHECKS
- Background checks, e.g., biometrics, watch lists...

CARD ISSUANCE
- Production of the national ID card

Figure 3.6 **Registration and card issuance process**

Applicants are supposed to come to any of the available registration centres with the supporting documents (application form,* passport and family book for citizens). The applicant goes through a three-stage (12 to 17 minutes) process. In the first office, the particulars completed on the application are scanned into the system through the 2D barcode on the form and verified against the online immigration system of the Ministry of Interior.

The applicant then moves to another station where his/her photo is taken, a signature is acquired using an electronic pad (which is then digitised into the system), and some supporting documents are scanned in. He then goes through the last station where his/her ten fingerprints are recorded (flat, rolled, palm and writers' palm).

The applicant is then given a receipt indicating the date he/she must return to collect his/her card, or that it will be sent to him/her through a registered courier.

The choice of card delivery is left to the applicant. Before the card is printed, there are other processes that are run at the back-end for further investigation. A biometric check is performed against civil and forensic fingerprint databases to ensure that the person has not been registered in the system previously, and is not wanted by the police authorities. In normal cases, cards are printed and distributed within 48 hours.

---

* The application form is also available for download from the Internet.

7. Lessons learned

This section provides a review of the current business operations in the UAE national ID card program and presents some of the identified areas for improvement. The proposed changes are seen as key factors to increase public acceptance and the project's chances of success.

7.1 The registration process

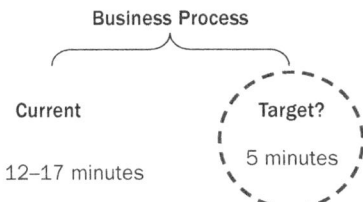

It was a concern from the early days of the project that the enrolment process involved many activities that required the project members to radically review and improve. The review exercise revealed that the average enrolment process could be completed in less than five minutes rather than the current 17 minutes without any big impact on the project objectives if only two enrolment stages were re-engineered: (1) the registration form and (2) the recording of biometrics.

7.1.1 The registration Form

The registration form throughout the project life-cycle went through many iterations in an attempt to reduce the amount of data needed for the enrolment. It started with an eight-page document, which was reduced to six, and then to four pages. The form, which was a pre-requisite to initiating the registration process, was viewed as follows:

- Too lengthy.
- Required considerable time to fill in.
- Some information was not readily available.
- Was sometimes filled incorrectly.
- Large number of resident applicants were illiterate.
- Considered to be the enrolment's bottle neck.

The reason for its design and the large amount of data required was to achieve the objective of producing statistics about the population of the country. The review process indicated that there was a mix-up between the two requirements of building a statistical database and of enrolling the whole population of the UAE and producing ID cards for them. This was a clear source of confusion among many members and stakeholders of the project when aiming to achieve these two objectives at the same time.

The recommendation from the review exercise was that the implementation of the project must take place in three stages as depicted in Figure 3.7. In the first stage, the project must attempt to (a) enrol the population for the new ID card with a minimal set of data as depicted in Figure 3.8 below. As only primary identification data will be required for first-time applicants, the application form was suggested to be eliminated and instead to make use of the existing electronic link with the Ministry of Interior's database to obtain and verify data.

Then stages two and three must run in parallel. In stage two, efforts must be directed towards promoting (and enforcing) the presentation of the new ID card for identity verification and as a pre-requisite to the most visited government services by the population.

Those organisations then need to maintain the new ID numbers in their databases, which should be used when moving to stage three of the strategy, and which requires the national ID database to interface and integrate with them. Provided the link is in place, a proper data warehouse can be built that is up-to-date and more reliable for statistical

Names	Name (Segmented) Gender, DoB, Marital St.	Name Gender, DoB, Martial St.
Passport	Nationality Passport No, Place of issue, Issue and Expiry dates Unified No Family ID, Book No	Nationality Passport No, Place of issue, Issue and Expiry dates Unified No, Sponsor Name Residency File No, Issue and Expiry dates
Address	Emirate City Mobile phone	Emirate City Mobile phone

Figure 3.7   Primary identification data captured for first enrolment

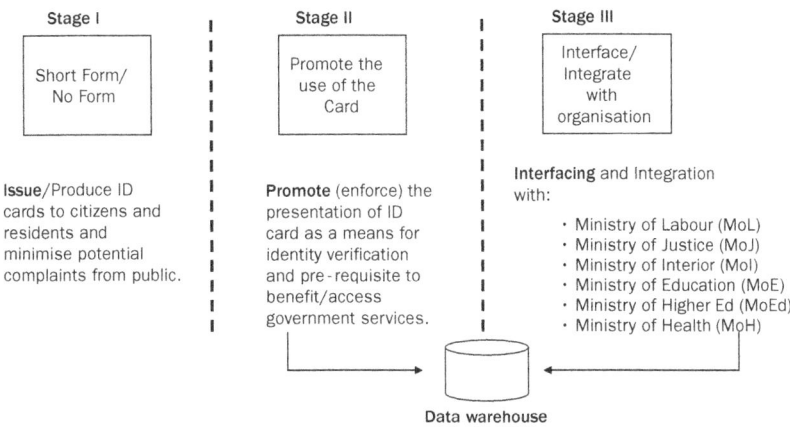

Data warehouse

Figure 3.8 Project implementation stages

reporting purposes as it will obtain information from primary and trusted sources.

7.1.2 The recording of biometrics

The enrolment process required the recording of all fingerprints (i.e., slaps, rolled prints, palm, writer's palms), a process taking around six to ten minutes to complete.

On the system, only rolled fingerprints were used for identification, while slaps and palms were stored for criminal search by the Ministry of the Interior.

The review recommendation suggested to only record flat prints and use smaller acquisition devices (just the slaps and the two thumbs), and to record other fingerprints at a later stage only if needed.

It was recommended that a second biometric must be introduced to complement the fingerprint biometrics. The second biometric was recommended to be more of a real time application that could be used in mass population areas such as airports. Both biometrics were seen as easy to operate and will cut processing time to less than 2 minutes.

7.1.3 Biometrics quality

Shortly after the introduction of the pilot phase of the programme it became abundantly clear that the quality of fingerprints taken by operators would have a determining effect on the classification, identification, and authentication of applicants. Apart from possible shortcomings in the operating system itself the percentages of failure to enrol (FTE), false rejection rate (FRR), and false acceptance rate (FAR) may increase dramatically if operators are not properly trained in taking fingerprints.

Failure to enrol due to operator failure may, for example, result in false demographic information. The lesson learned was that a very high premise should be placed on a comprehensive operator training programme. It was also realised that results of biometrics should be closely monitored to determine the performance of the system in terms of the quality of these results.

A clear indication of unacceptable system performance would be if a too long hit-list is required to identity hits, or if the real hit constantly appears very low on this list. While it can be argued that the list can be shortened by fine-tuning the applicable thresholds, it will then mean that real hits that appear low on the list will not be identified if the system's performance is not improved.

There is obviously a very close relationship between the quality of fingerprints taken and the performance of the system. It was, however, realised that the introduction of a second biometric would complement the fingerprint biometric and balance any shortcomings.

The re-engineering of the above two enrolment processes provided a saving of office space, equipment and staff required for enrolment, as the original process was divided into three stages as depicted in Figure 3.9.

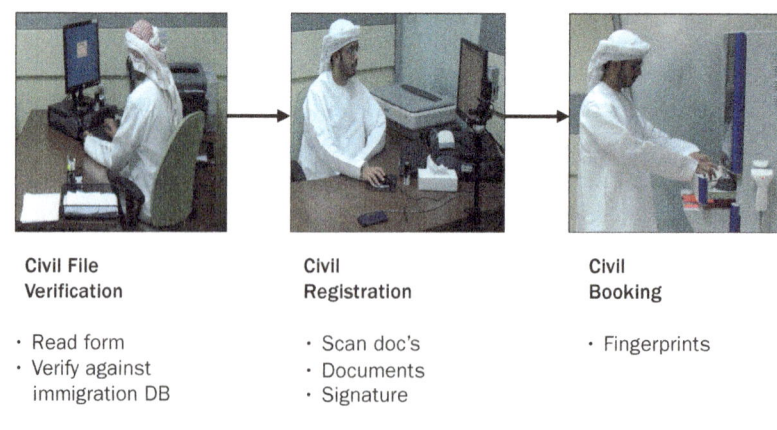

Civil File
Verification

· Read form
· Verify against
  immigration DB

Civil
Registration

· Scan doc's
· Documents
· Signature

Civil
Booking

· Fingerprints

Figure 3.9   Registration procedures

The reason for the three office designs was mainly to separate duties, manage the daily in-flow of applicants, and shorten the waiting time.

It was therefore recommended to improve the system by carrying out the enrolment process through a single workstation. This was seen to radically support the enrolment process, in which smaller devices could be used to carry out the registration and improve the portability of the system for wider deployment in areas such as setting up permanent and temporary registration offices in traffic departments, immigration, municipalities, schools, companies with a large number of staff members, and so on.

7.2 Scope creep

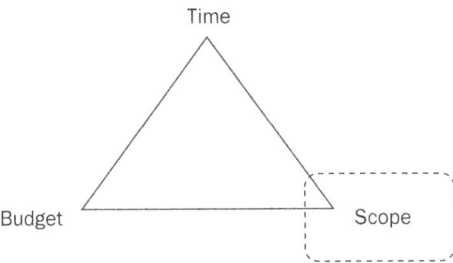

'Scope creep' is a term used to describe the process by which users discover and introduce new requirements that are not part of the initial planning and scope of the project. As widely quoted in the literature, many doubt a limited and specified scope for national ID programmes, as the nature and high cost of such projects are likely to yield and encourage the expansion of its functions (5, 6, 7, 8).

The project management literature also indicates that coping with changes and changing priorities are the most important single problems facing the project management function (9, 10, 11). Indeed, changing targets all the time would obviously take any project nowhere.

In the UAE ID programme, project performance was monitored and measured regularly to identify variances from the project plan. It was supported by a formal and well-defined process to control and manage the changes being requested to the project scope and objectives during the project lifecycle. See for example Figure 3.10 that depicts an example of one of the changes in control policies.

Despite all the efforts put into managing and controlling change, the changes introduced and processed dragged the project schedule far away from the original target date. It was common during the different implementation stages of the programme to change the scope to either

Type of Change			
Technical nature (No financial impact)	Technical nature (With financial impact)	Contract Scope (No financial impact)	Contract scope (Financial Impact)
Steering Committee — Accept or Reject	Analyse and Recommend	Accept or Reject	Analyse and Recommend
Change Control Board	Reject or Recommend to Project Sponsor		Reject or Recommend to Project Sponsor
Project Sponsor	Approve or Reject		Approve or Reject

Figure 3.10 Change in control authorities

add new, or change, agreed functions, which obviously had a severe impact on the implementation of the project's plan and budget. Examples of such changes included the shifting from centralised to decentralised card-printing, upgrading the card technology, changing card design and displayed data, upgrading database technology, and so on.

Another example of such changes to the project scope was related to the perception of the multi-purpose functionality. Several attempts were made from certain key stakeholders of the project to expand the card applications during the execution phase, which had a severe impact on the project's progress.

It took some time for them to realise the importance of limiting the purpose of the card as an identity document in the first phase of the project and that efforts must rather be concentrated on the enrolment of the population and the issue of the new ID card.

Indeed, a multi-purpose card was one of the objectives of the project, but not in the way it was comprehended. The multi-purpose term stated in the objectives was used to explain that the card could replace other identity documents when it came to the verification of identities. Since the card was obligatory for the total population of the UAE, the provision was that the new ID card could replace such cards if the other organisations used the new ID number in their databases as one with which to retrieve individual records.

The management of scope in the UAE ID card was clearly one of the biggest challenges, which required the project core team to spend a lot of time and effort in clarifying the feasibility of such actions to the upper management. Finding the right communication approach was key to managing scope creep.

7.3 *Too much security*

It was during this evaluation phase of the project (discussed next) as well as the experience gained after the introduction of the pilot phase that it was realised that far too much emphasis was being placed on security issues which rendered the operating system a closed one which required costly and intensive efforts to affect even simple changes. Needless to say, the required changes to the system became a very cumbersome task with unacceptable costs and time frames associated with each change.

This resulted in a strained relationship with the vendor as these delays were perceived by the client as reflecting on a possible inability or lack of co-operation on the part of the vendor. It was later realised that the security during the enrolment process and security around the information included in the card itself should not be to such an extent that it placed a stranglehold on the flexibility to change and the user-friendliness of the system.

While security features built into the card could be as inclusive as required, the personal information of the applicant stored in the security portion of the card should be protected but should be freely available to authorised users.

7.4 *Evaluation framework*

Since the UAE national ID system was provisioned to become the most critical system in the country as the main hub for population identity cross-checking and service eligibility (i.e., online with 24/7 availability requirement), it was important that the overall system went through a thorough quality evaluation.

As widely quoted in the literature, one of the principal causes of an information system failure is when the designed system fails to record the business requirements or improve the organisational performance (see for example: [12, 13, 14, 15, 16]). Figure 3.11 shows an example of how a user's requirements might be interpreted during the lifecycle of a project, which is not really far from being true in many of the IS projects implemented around the world.

During the UAE programme implementation there was no clear communication of the development standard followed by the vendor, which created confusion among the project team members when it came to individual delivery, as well as the final acceptance of the system.

In general, the project team, with the workload and responsibilities put on them, seemed to be snowed under and to have differing ideas

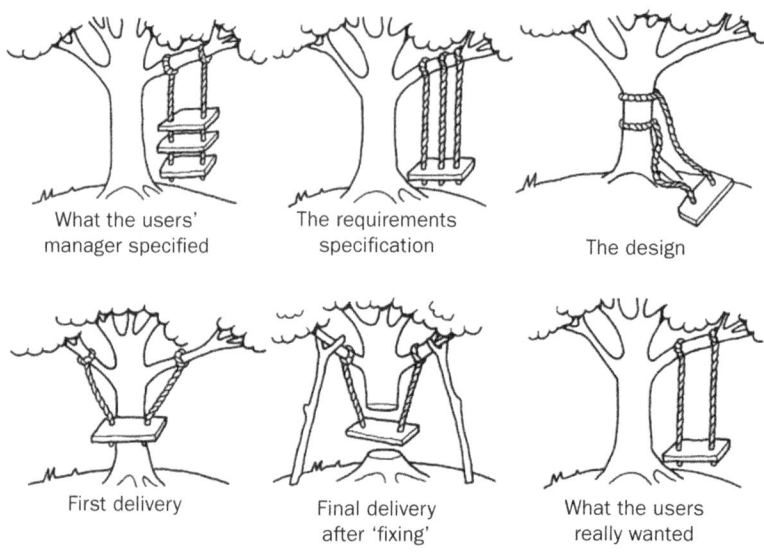

What the users' manager specified	The requirements specification	The design

First delivery	Final delivery after 'fixing'	What the users really wanted

Figure 3.11   An example of varying interpretations of user requirements

and visions of how things should be done. Everybody wanted the project to be concluded as quickly as possible and was seemingly very impressed with the work produced by the vendor.

During the very late stages of the project, the core project team employed ISO 9126 standard for the purposes of software quality and the overall evaluation of the system's architecture (see also Figure 3.12.).

The evaluation study contributed significantly to identifying many of the system's deficiencies that required the vendor to address prior to the final acceptance and handover of the system. Besides, the use of a quality framework provided a very useful and supportive methodological approach for going about software quality assessment. It acted as a comprehensive analytical tool and provided a more thorough view of the system's strengths and weaknesses.

It addressed a wide range of quality characteristics of the software products and processes enabling a better description of the software quality and importance. Arguably, if used as a guide in an early stage of the project it could have provided a sound basis for informed and rational decision-making which could have contributed significantly to the delivery of a system which properly addresses user requirements.

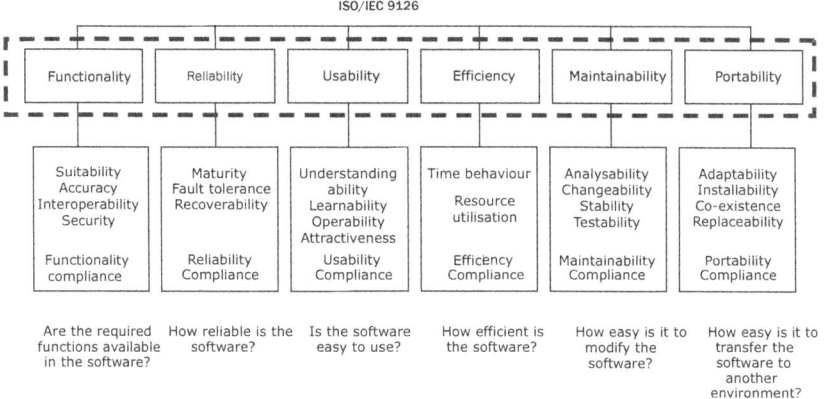

Figure 3.12 ISO-IEC 9126 quality model

8. Conclusion

Not much has been written about the national ID system's implementation from a practitioners' point of view. The literature was found to be full of articles from the private sector discussing advanced technologies and what can be achieved through them. This article adds to the current body of knowledge in this field, and is believed to assist in widening the circle of those who are of the same opinion.

As indicated in this paper, much could indeed be achieved through national ID programs. However, agreed ideas and proper planning is essential to succeed with such projects. It took quite some time before the team started to agree on the fact that the first phase of the project should focus on the enrolment of the whole population and the issue of the new ID card.

National ID programs, because of their nature, are perceived to invest quite significantly in technology upgrades and Research and Development departments for future development in areas related to identity management, and to keep up with developments and standards in the ID industry. It was learned that new functionalities and upgrades must be thoroughly studied and researched to eliminate unnecessary changes during the execution phase.

As it is becoming the trend, such programs need to put a lot of effort in to promoting e-identity and e-verification services using the new ID card. It would be interesting to measure the impact of national ID programmes on the overall government economy, as it would obviously promote electronic transactions and would also encourage government organisations to streamline their operations and make use of the secure e-verification infrastructure that it will provide.

8.1. Further research

Further research and practical work in this field may further contribute to the body of knowledge. Areas in which further research may yield valuable insights and better understanding are:

1. Similar studies of national ID implementation that show the appropriateness of the items presented in the lessons learned, and the degree to which they can support similar initiatives.

2. An investigation of the suitability of quality models in national ID programmes and projects of a similar nature and their impact on the project success rate.

Acknowledgment

The author would like to thank Mr. Gawie von Wielligh for his review feedback that improved the overall structure and quality of this paper.

References

1. Goldman, L. (2004) 'Chips with everything'. *Occupational Health* 56 (4): 12.
2. Knight, H. (2003) 'Anti-Fraud: The road to a national ID card is paved with pitfalls'. *The Engineer* December: 25.
3. Matthews, W. (2002) 'ID card plan assailed'. *Federal Computer Week* 16 (4): 16.
4. Michael, H. (2006) 'Identity cards: Why bother?' *The Economist* 335 (7916): 50.
5. Clarke, R. A. (2002) 'The Resistible Rise of the National Personal Data System'. *Software Low Journal* 5 (1): 33–36.
6. Clarke, R .A. (1994) 'Human Identification in Information Systems: Management Challenges and Public Policy Issues' Australia: The Australian National University. Available at: *http://www.anu.edu.au/people/Roger.Clarke/DV/HumanID.html#PPI*

7. Fontana, J. (2003) 'A national Identity Card for Canada'. Canada: House of Commons. Available at: *http://www.parl.gc.ca*
8. Froomkin, A. M. (2002) 'The Uneasy Case for National ID Cards as a Means to Enhance Privacy'. USA: University of Miami School of Law. Available at: *http://www.la-w.tm*
9. Heerkens, G. R. (2005), 'Project management: 24 lessons to help you master any project'. New York: McGraw-Hill.
10. Meredith, J. R. and Mantel, S. J. (2003) 'Project management: a managerial approach' 2nd edition. New York: John Wiley and Sons.
11. Reiss, R. G. and Spon, F. N. (1995) 'Project Management Demystified: Tools and Techniques'. New York: McGraw Hill.
12. Avison, D. E. and Wood-Harper, A. T. (1990) 'MultiView – an exploration in information systems development'. USA: McGraw Hill.
13. Bocij, P., Chaffey, D., Greasley, A. and Hickie, S. (2003) 'Business Information Systems: Technology, Development and Management for the e-business' (2nd edition). New York: Prentice Hall.
14. Crain, W. (1992) 'Theories of Development: Concepts and applications' (3rd edition). New Jersey: Prentice Hall International.
15. Curtis, G. (1998) 'Business Information Systems: Analysis, Design, and Practice' (3rd edition). USA: Addison-Wesley.
16. Harry, M. 'Information Systems in Business' (2nd edition). Great Britain: British Library Cataloguing in Publication D.

4

Using quality models to evaluate large IT systems

Ali M. Al-Khouri

Abstract: This article presents findings from the evaluation study carried out to review the UAE national ID card software. The article consults the relevant literature to explain many of the concepts and frameworks explained herein. The findings of the evaluation work that was primarily based on the ISO 9126 standard for system quality measurement highlighted many practical areas that, if taken into account, is argued to increase the chances of success of similar system implementation projects.

Key words: *National ID system, software quality, ISO 9126.*

1. Introduction

The United Arab Emirates (UAE) have recently initiated a national ID scheme that encompasses very modern and sophisticated technologies. The goals and objectives of the UAE national ID card programme go far beyond introducing a new ID card document and homeland security (1).

To increase its success, the government is pushing for many innovative applications to explore 'what can be done with the card'. Examples of such possible applications of the card range from using it as a physical identity document, to proving identity, to linking it to a wide range of government services, with the vision of replacing all existing identity documents (e.g., driving license, labour card, health card and so on). with this new initiative.

From these perspectives, it becomes critical that such systems maintain a high level of quality. Quality models can play a good role as useful tools for quality requirements in engineering, as well as for quality evaluation, since they define how quality can be measured and specified (2). In fact, the literature reveals that the use of quality frameworks and models may well contribute to project success, as it enables the early detection and addressing of risks and issues of concern at an early stage of the project (see for example 3, 4, 5).

This article attempts to provide a short evaluation of the population register software (referred to in this article as PRIDC – population register and ID card) implemented as part of the national ID card project in the UAE to pinpoint areas of possible improvement.

The article is structured as follows. The first section provides brief background information about the concept of software quality and measurement standards, with a focus on the ISO 9126 framework. The next section presents the methods employed to obtain data based on which the system was evaluated. The next few sections provide an overview of the PRIDC system, its components, its developing lifecycle approach, results obtained from the previous tests, and mapping these latter sets of data to ISO 9126 quality attributes. The paper is then concluded with some reflections on the areas that need to be considered when pursing similar evaluation studies with a focus on national ID systems.

2. Software quality

It is becoming a common trend for IT projects to fail. The rate of failure in government projects is far higher than those in the private industry. One of the main causes for such failures was widely quoted in the literature to be related to poor user requirements resulting in a system that does not deliver what was expected from it (see also the statistics presented in Figure 4.1 from the recent Standish Group study).

The CHAOS survey of 8000+ projects found that of the eight main reasons given for project failures, five requirements are related. Getting the requirements right is probably the single most important thing that can be done to achieve customer satisfaction. Figure 4.2 depicts further reasons for such failures (6). Many of these failures could have been prevented with requirements verification and the adoption of quality assurance frameworks (4, 7).

In general terms, there are three different approaches to system quality assurance:

Standish Study Results:

51%	Failed projects
31%	Partially successful

Failure causes:

13.1%	Incomplete requirements
12.4%	Lack of user involvement
10.6%	Inadequate resources
9.9%	Unrealistic user expectations
9.3%	Lack of management support
8.7 %	Requirements keep changing
8.1%	Inadequate planning
7.5 %	System no longer needed

Figure 4.1   Standish group study results

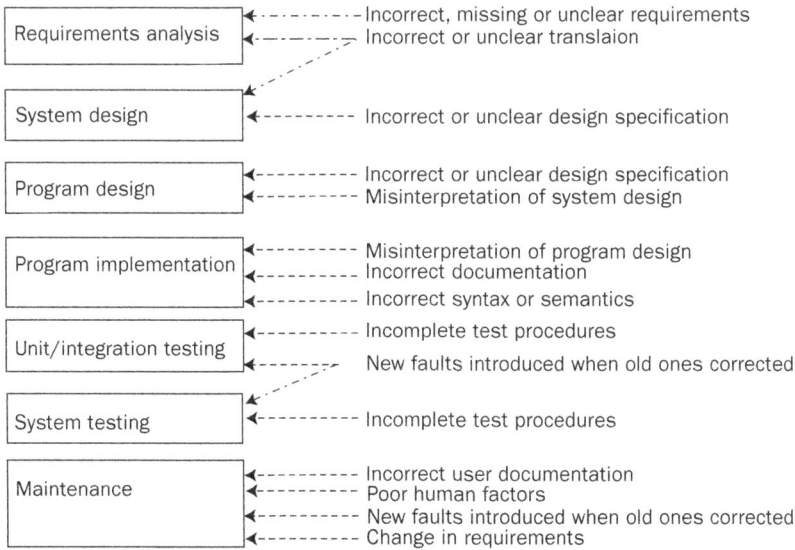

Figure 4.2   Causes of faults during development

1. Product Certification

An independent party (or a QA company) conduct a limited exercise in verification, validation and/or test of the software components.

2. Process audit

An independent party conduct and assess the development process used to design, build and deliver the software components.

3. User satisfaction

Analysis of the actual behaviour of the software.

Since the objective of the evaluation study in this article is to judge whether the given implemented system has met the requirement of product quality, the third approach was defined as the boundaries for the evaluation taking place in this study.

2.1 Quality measurement standards

Software quality assessment is attracting great attention as the global drive for systemic quality assurance continues to gather momentum, for example, pressures of consolidations, mergers, downsizing, and the emergence of new technologies (8).

The preliminary work conducted in the field of software quality assessment was done by B. Boehm and associates at TRW (9) and incorporated by McCall and others in the Rome Air Development Center (RADC) report (10).

The quality models at the time focused on the final product and on the identification of the key attributes of quality from the user's point of view. The assessment framework was later improved; consisting of quality attributes related to quality factors, which were deconstructed into particular quality criteria and led to quality measures (see Figure 4.3).

Attempted standardisation work over the intervening years resulted in the Software Product Evaluation Standard, ISO-9126 (ISO/IEC, 1991). This model was fairly closely modelled on the original Boehm structure, with six primary quality attributes that were subdivided into 27 sub-characteristics as illustrated in Figure 4.4.

However, the standard was criticised for providing very general quality models and guidelines, and for being very difficult to apply to specific domains such as components and CBSD (11, 12). However, this is

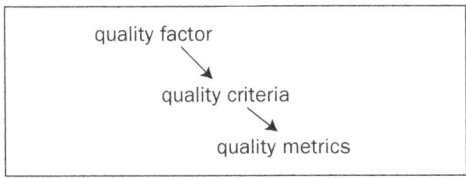

Figure 4.3   Boehm quality model

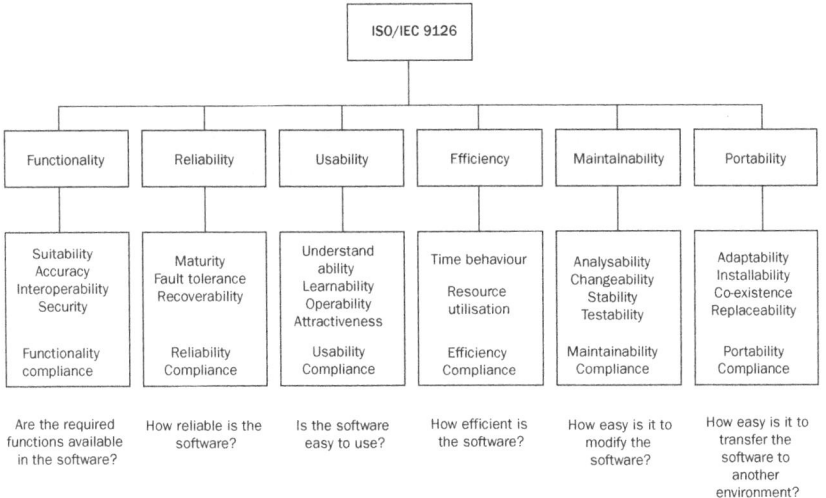

Figure 4.4   ISO/IEC 9126 standards characteristics

believed by others to be in fact one of its strengths as it is more adaptable and can be used across many systems (13, 14). To solve this problem ISO/IEC 9126 has been revised to include a new quality model which distinguishes between three different approaches to product quality:

1. External Quality metrics – ISO TR 9126 – 1: a result of the combined behaviour of the software and the computer system and can be used to validate the internal quality of the software.

2. Internal Quality metrics – ISO TR 9126 – 3: a quantitative scale and measurement method, which can be used for measuring an attribute or characteristic of a software product.

3. Quality in use metrics – ISO TR 9126 – 4: is the effectiveness, productivity and satisfaction of the user when carrying out representative tasks in a realistic working environment. It can be used to measure the degree of excellence, and can be used to validate the extent to which the software meets user needs. Figure 4.5 depicts the relationship between these approaches.

In brief, internal metrics measure the software itself, external metrics measure the behaviour of the computer-based system that includes the software, and quality in use metrics measure the effects of using the software in a specific context of use.

Appropriate internal attributes of the software are prerequisites for achieving the required external behaviour, whereas external behaviour is a prerequisite for achieving quality in use (see also Figure 4.6).

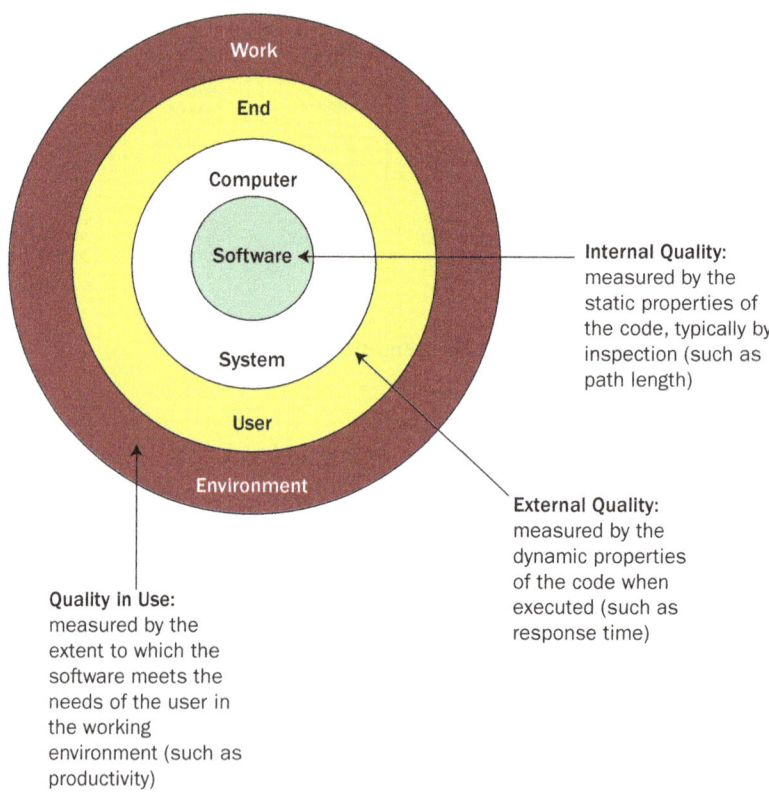

Figure 4.5 Relationship between internal quality, external quality and quality in use

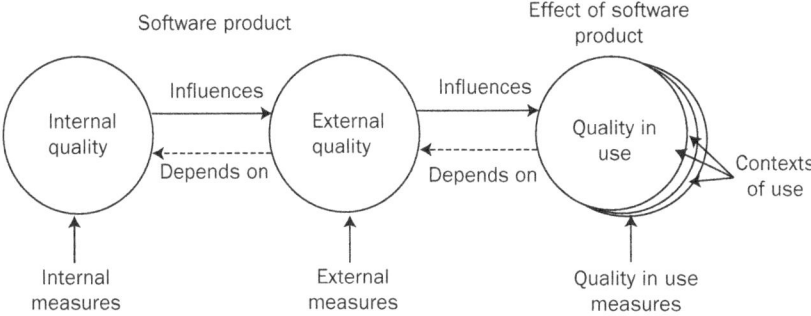

Figure 4.6 Approaches to software product quality

It is also worth mentioning that a new project was launched called SQuaRE – Software Product Quality Requirements and Evaluation (ISO/ IEC 25000, 2005) – to replace the above but to follow the same general concepts of 9126 standard (see also Figure 4.7).

Nonetheless, research and practical work shows that the assessment of the quality of a software component is in general a very broad and ambitious goal (11).

Recent research also shows that these characteristics and sub-characteristics cover a wide spectrum of system features and represent a detailed model for evaluating any software system as Abran et al. (15) explain:

> …ISO 9126 series of standards …. even though it is not exhaustive, this series constitutes the most extensive software quality model developed to date. The approach of its quality model… is to represent quality as a whole set of characteristics… This ISO standard includes the user's view and introduces the concept of quality in use.

The five divisions of SQuaRE standard:

(1) *Quality management division (ISO 2500n)*
(2) *Quality model division (ISO 2501n)*
(3) *Quality measurement division (ISO 2502n)*
(4) *Quality requirements division (ISO 2503n)*
(5) *Quality evaluation division (ISO 2504n)*

Figure 4.7 SQuaRE standard

3. Methodology

'If you chase two rabbits! both will escape.' Chinese Proverb

ISO 9126 quality characteristics and sub-characteristics were used to evaluate the national ID system. In this investigation several evaluation methods were employed. The following are the prime sources of information for the evaluation study:

1. Information gathered from the test sessions took place during the acceptance of the project deliverables.

2. Observation of the system environment (both at the central operational and registration centres);

3. by means of recording the author's own experience as the Director of the Central Operations sector, and head of the technical committee overseeing the implementation of the program.

In general, the evaluation was qualitative in nature. In carrying out the evaluation and recording the findings, the PRIDC system went through two types of testing; functional and technical.

3.1 Functional testing

This is an application level testing from a business and operational perspective. It is conducted on a complete, integrated system to evaluate the system's compliance with its specified requirements. Often called black box testing, this type of test is generally performed by QA analysts who are concerned about the predictability of the end user experience. During the deliverables acceptance, the national ID system was tested with black box testing procedures (which focus on testing functional requirements and do not explicitly use knowledge of the internal structure) as per the test plan designed by the solution provider. No change was allowed by the vendor to the test plan as they wanted to narrow down the scope of testing, and limit it to the test cases developed by them.

3.2 Technical testing

This is the system level testing. It tests the systems which are support or enable the running of Functional Applications. With a general perception of QA, the COTS are not seen to be required to be tested, but they need to be audited for the configuration and deployment set-up. Generally, white box testing (also called glass, structural, open box or clear box testing) was considered here by the technical team to test the design of the system that should allow a peek inside the 'box', as this approach focuses specifically on using internal knowledge of the software to guide the selection of test data.

White box testing requires the source code to be produced before the tests can be planned and is much more laborious in the determination of suitable input data and the determination if the software is, or is not correct. It is worth mentioning that a failure of a white box test may result in a change which requires all black box testing to be repeated and the re-determination of the white box paths. For this obvious reason there was always negligence from the vendor to initiate white box testing. It must also be heeded that neither black nor white box testing can guarantee that the complete specifications have been implemented and all parts of the implementation have been tested.

To fully test a software product, both black and white box testing are required. While black box testing was limited by the test plan documents provided by the vendor, the white box testing was not possible since the source code was still not handed over to the client at the time of writing this study. However, all the architectural components of the national ID sub-systems which were selected and assembled from the COTS were assessed and audited to their configuration and deployment set up. Having addressed the evaluation methods, the following sections describe the details of the work carried out in this study.

4. The PRIDC system as a component-based system

For more than a decade good software development practice has been based on a divide and conquer approach to software design and implementation. Whether they are called modules, packages, units, or computer software configuration items, the approach has been to decompose a software system into manageable components based on maximizing cohesion within a component and minimizing coupling among components.

(Brown and Wallnau, 1996: 414)

4.1 What is component-based software (CBD)?

Component-based software development (CBD) is an emerging discipline that promises to take software engineering into a new era (16). Building on the achievements of object-oriented software construction, it aims to deliver software engineering from a cottage industry into an industrial age for Information Technology, wherein software can be assembled from components, in the same manner that hardware systems are currently constructed from parts (ibid).

Component-based software development (CBSD) shifts the development emphasis from programming software to composing software systems, as it embodies the 'buy, don't build' philosophy espoused by (17). See also Figure 4.8. The concept is also referred to in the current literature as component-based software engineering (CBSE) (18, 19). It principally focuses on building large software systems by integrating different software components and enhancing the overall flexibility and maintainability of the systems.

If implemented appropriately, the approach is argued to have the potential to reduce software development costs, assemble systems rapidly, and reduce the spiraling maintenance burden associated with the support and upgrade of large systems (20).

(21) defines component-based software development as an approach 'based on the idea to develop software systems by selecting appropriate off-the-self components and then to assemble them with a well-defined software architecture'. They state that a component has three main features:

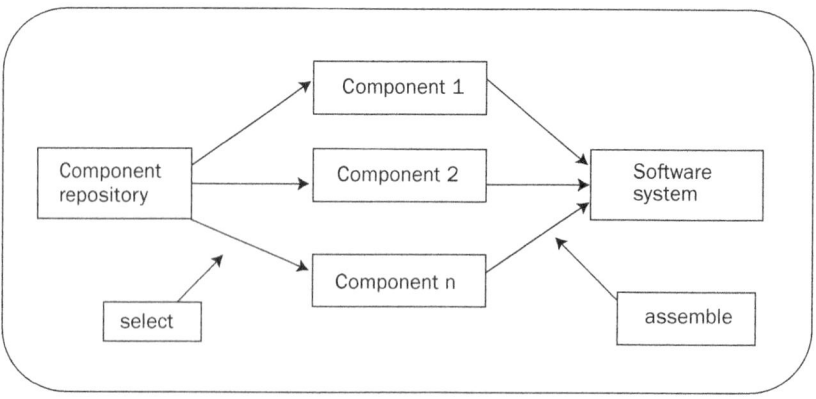

Figure 4.8　　Component-based software development

1. It is an independent and replaceable part of a system that fulfils a clear functions;

2. It works in the context of well-defined architecture.

3. It communicates with other components by its interface.

According to (22), two main advances are raising the profile of software components as the basic building blocks of software – see also: (16, 23, 24, 25, 26, 27):

1. The object-oriented development approach which is based on the development of an application system through the extension of existing libraries of self-contained operating units, and

2. the economic reality that large-scale software development must take greater advantage of existing commercial software, reducing the amount of new code that is required for each application.

A component-based development approach introduces fundamental changes in the way systems are acquired, integrated, deployed and evolved. Unlike the classic waterfall approach to software development, component-based systems are designed by examining existing components to see how they meet system requirements, followed by an iterative process of refining requirements to integrate with the existing components to provide the necessary functionality (22).

4.2 The component-based software development lifecycle

The component lifecycle is similar to the lifecycle of typical applications, except in the implementation and acquisition phases where the two lifecycles differ. The lifecycle of component-based software systems can be summarised as follows:

1. Requirement analysis: a process of defining and understanding the activities that the information system is meant to support;

2. software architecture design: a process of developing detailed descriptions for the information system;

3. component identification and customisation (implementation): a process of formalising the design in an executable way by acquiring complete applications or components through purchase, outsourcing, in-house development, component-leasing, and so on;

4. system integration: a process of adjusting the system to fit the existing information system architecture. This can include tasks such as adjusting components and applications to their specific software surroundings;

5. system testing: a process of identifying and eliminating non-desirable effects and errors and to verify the information system. This can include both user-acceptance and application integration tests;

6. software maintenance: a process of keeping the integrated information system up and running. This can include tasks such as upgrading and replacing applications and components in the information system. It also includes performing consecutive revisions of the integrated information system.

Having briefly highlighted some background information about the concept of component-based development and lifecycles the next section takes a snapshot of the PRIDC system and maps onto component-based software.

4.3 The PRIDC system development life cycle

Broadly speaking, the development of the PRIDC system in general can be described to have incorporated the following two approaches:

1. The development of a uniquely tailored information system (population register) to enable the registration of the population into the system in accordance to the pre-defined business requirements, and

2. the integration of several application/hardware package to achieve the desired functionality requirements e.g., biometrics, PKI, smart cards.

For the purpose of benchmarking the PRIDC system development lifecycle, a framework proposed by (28) for the quality assurance of a component-based software development paradigm has been adopted in this study. The framework contains eight phases relating to components and systems that provide better control over the quality of software development activities and processes:

1. Component requirement analysis.
2. Component development.
3. Component certification.
4. Component customisation.
5. System architecture design.
6. System integration.

7. System testing.

8. System maintenance.

The details of this benchmarking are presented in the following section.

4.4 A comparison of the PRIDC system with ISO standard

The PRIDC system lifecycle is currently based on project implementation phases. The project implementation is executed as framed in the contract. A comparative study of the PRIDC system lifecycle with the ISO 12207 standard can be presented as below:

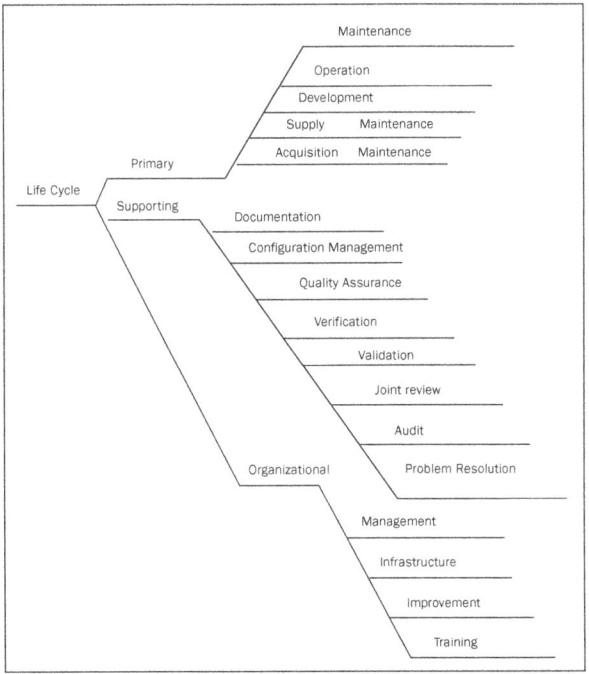

Figure 4.9   The ISO 12207 standard*

---

*The ISO standard for system implementation.

In 1987 ISO and IEC (International Electrotechnical Commission) established a joint Technical Committee (JCT1) on Information Technology. In June 1989, the JCT1 initiated the development of ISO/IEC 12207 on software lifecycle processes to fill a critical need. The ISO was published 1 August 1995.

Table 4.1 Comparison of the PRIDC system with the ISO 12207 standard

No	ISO 12207	PRIDC System Life cycle
1	**Primary Life Cycle Processes**	
1.1	Acquisition Process	All lots of Category A.
1.2	Supply Process	All lots of Category D.
1.3	Development Process	All lot of Category B.
1.3.1	Process Implementation	All lots of Category C.
1.3.2	System Requirement Analysis	All lots of Category A.
1.3.3	System Architectural Design	All lots of Category B.
1.3.4	Software Requirement Analysis	The solution provider internal process.
1.3.5	Software Architectural Design	The solution provider internal process.
1.3.6	Software Detail Design	The solution provider internal process.
1.3.7	Software Coding and Testing	The solution provider internal process. EIDA had done few of such testing.
1.3.8	Software Integration	The solution provider internal process.
1.3.9	Software Qualification Testing	The solution provider internal process.
1.3.10	System Integration	The solution provider internal process.
1.3.11	Software Qualification Testing	The solution provider internal process.
1.3.12	Software Installation	The solution provider internal process.
1.3.13	Software Acceptance Support	Lot 12 and Lot3 testing
4	Operation Process	Need to define.
5	Maintenance Process	Need to define.
6	**Supporting Life Cycle Processes**	
6.1	Documentation Process	Done as a part Project deliverable.
6.2	Configuration Management Process	Done as a part Project deliverable.

Table 4.1	Comparison of the PRIDC system with the ISO standard (*Cont'd*)	
6.3	Quality assurance process	Not done as a part of Project contract.
6.4	Verification process	Can be considered, Lot 3 system compliance test.
6.5	Validation process	Can be considered, Lot 12 system compliance test.
6.6	Joint review process	Can be considered – program management meeting.
6.7	Audit process	Need to perform.
6.8	Problem resolution process	Need to perform.
7	**Organisational life cycle processes**	
7.1	Management process	EIDA needs to perform.
7.2	Infrastructure process	EIDA needs to perform.
7.3	Improvement process	EIDA needs to perform.
7.4	Training process	Done as a part of Lot 3 – Admin Training.

4.5 ISO 9126 and PRIDC mapping

The following is a summary of the evaluation results as per the ISO 9126 quality attributes.

Functionality	The degree of existence of a set of functions that satisfy stakeholder/business implied needs and their properties. Overall, in terms of number of changes requested on the system, there were more than 213 modification items in the form of 53 change requests (23 major changes) passed to the vendor to implement. These were the first functional test results with the first version of the PRIDC system. This is a significant number of modifications and it clearly implies that there was a big gap during the system requirement analysis and recording phase.

(continued)

Suitability	Can software perform the tasks required? The degree of the presence of a set of functions for specified tasks (fitness for purpose).	Checked against specifications and feedback from registration centres.
Accuracy	Is the result as expected? The degree of provision of right or agreed results or effects.	Checked against specifications. Test cases were developed by the test team of the vendor. Besides, there were many other cases that were not tested for accuracy but encountered later after the release of the software.
Interoperability	Can the system interact with another system? The degree to which the software is able to interact with specified systems (i.e. physical devices).	Checked against specifications. However, the system was designed to be closed, as inter-operability with future systems was seen to be of big concern.
Security	Does the software prevent unauthorised access? A set of regulations for maintaining a certain level of security; degree to which the software is able to prevent unauthorised access, whether accidental or deliberate, to programs and data (i.e. log in functions, encryption of personal data and so on).	Checked against specifications and in accordance with the Information Security Policy. The PRIDC system is a critical system for the country, thus important security features were incorporated into the system to ensure high confidentiality, integrity and authenticity of the data. The security is built around the following main rules: ■ Strong authentication of the operators (each end user will use both password and fingerprint to log onto the system).

		▪ Network security using Virtual Private Network (VPN) + Demilitarised Zone (DMZ) and Secure Socket Layer (SSL) over Hyper Text Transfer Protocol (HTTP). ▪ Strong physical protection of the Central, Disaster Recovery and Service Points Local Area Networks (LAN) The security scheme was implemented at 4 levels: 1) Application level, 2) Network level, 3) System level, 4) Physical level. The security features carried out at each of the above levels included a wide range of advanced international security standards and measures: X509 V3 certificates, X500 directory, LDAP V2 and V3, DES, 3xDES, RC2, RC4, AES ciphering algorithms (used by CISCO VPN), RSA (PKCS#1) signature algorithms, MD2, MD5, SHA1, Diffie-Hellman and RSA key exchange algorithms, pkcs#12, pkcs#7, pkcs#10, IPsec.
Compliance	The degree to which the software adheres to application-related standards or conventions or regulations in laws and similar prescriptions.	Checked against specifications.

(continued)

Reliability	The capability of the software to maintain its level of performance under stated conditions for a stated period of time (This is assessed based on the number of failures encountered per release).	
Maturity	Have most of the faults in the software been eliminated over time? The frequency of failure by faults in the software.	If we look at the number of sub-versions released of the PRIDC system (ie., ver 1.5, ver 1.6 and ver 3.0) The evolution of these versions was unplanned (i.e., previously not specified) which signifies the immaturity of the system in terms of business requirements and needs. At the time of carrying out this evaluation, the software was still seen to require further modifications before the system could be finally accepted.
Fault tolerance	Is the software capable of handling errors? The ability to maintain a specified level of performance in cases of software faults or of infringement of its specified interface is the property that enables a system to continue operating properly in the event of the failure of some of its components.	Although the system had a centralised architecture, it allowed the different systems to continue operating in the cases of failure of the central system through replication and redundant systems.
Recoverability	Can the software resume working and restore lost data after failure? The capability of software to re-establish its level of performance and recover the data directly affected in case of a failure.	Databases were continuously replicated on the Disaster Recovery site. The system insured that no more than one hour of work would be lost following a database crash/failure. However, in case of a major disaster that would lead to the loss of the operational capacity of the main data centre, the PRIDC system was planned to be restarted within 24 hours.

Usability	the effort needed for the use by a stated or implied set of users.	
Understandability	Does the user comprehend how to use the system easily? Evaluation of the attributes of software that bear on the users' efforts to recognise the underlying concept of the software. This effort could be decreased by the existence of demonstrations.	Usability testing uncovered many difficulties, such as operators having difficulty understanding the system interface, business logic and processes. With the lack of an online help function, the GUI interface of the system did not seem to follow any clear standard, as operators started guessing what different buttons may mean. For example, two registration centres' operators deleted the files of all registered applicants on one day when they pressed the button 'Abort' to cancel an operation, where the system was performing the action of 'Delete' with the 'Abort' button. In general, Interface functions (e.g., menus, controls) were not easy to understand.
Learnability	Can the user learn to use the system easily? Evaluation of the attributes of software that bear on the users' effort to learn how to use the software	User documentation and help were not complete at the time of carrying out this evaluation. The system was not easy to learn as users had to repeat the training sessions many times as the cases of data entry errors were raised when post-audit procedures for a data quality check were implemented.
Operability	Can the user use the system without much effort? Evaluation of the attributes of the software that bear on the users' effort for	The interface actions and elements were sometimes found to be inconsistent – error messages were not clear and led to more confusion and resulted in operators

(continued)

	operation control (e.g. function keys, mouse support, shortcuts and so on).	guessing and attempting to rectify problems which in turn led to deeper problems as the system was not designed to handle cases of user error (i.e., to handling unexceptional errors). Some important functions such as deletion was being performed with a prompt for confirmation.
Attractiveness	Does the interface look good? Does it evaluate how attractive is the interface to the user?	The system design and screen layout and colour was not so appealing
Efficiency	Have the functions been optimised for speed? Have repeatedly used blocks of code been formed into sub-routines?	
Time behaviour	How quickly does the system respond? Evaluating the time it takes for an operation to complete the software's response and processing times and rates in performing its function.	To be checked against specification. However, the full testing of this characteristic was not possible at the time of carrying out this study since the daily enrolment was around 1200 people subsequently there were the same figures for the card production. From a database capacity viewpoint, the PRIDC system was big enough to manage records of five million people. Whereas the throughput of the system was as follows: ▪ It allows for up to 7,000 enrolments per day. ▪ It is able to produce up to 7,000 ID Cards per day. ▪ The Biometric Sub-System is able to perform up to 7,000 person identification (TP/TP) searches per day.

		The processing operations were designed as follows: ■ New enrolment: 20 minutes ■ Card collection: 3.5 minutes ■ Card renewal: eight minutes ■ PR Functions: 8.5 minutes ■ Civil investigation: 11 minutes ■ Biometric subsystem: Within 22 hours
Resource utilisation	Does the system utilise resources efficiently? Is the process of making code as efficient as possible; that is, the amount resources and the duration of such use in performing the software's functions.	This task was not possible at the time of carrying out the evaluation, since the source code was still not handed over to the client.
Maintainability	The effort needed to make specified modifications	
Analysability	Can faults be easily diagnosed? The effort needed for diagnosis of inefficiencies or causes of failure or for identification of parts to be modified.	During system installation and with the release of the software (also during business operations), undocumented defects and deficiencies were discovered by the users of the software. These encountered faults were very difficult to analyse and diagnose even by the vendor technical team, and so encountered software inefficiencies usually took a long time to fix, as problems were usually passed to the development team in France for investigation and response.

(continued)

Changeability	Can the software be easily modified? Changeability is the effort needed for modification, removal or for environmental change.	The system architecture was so complex, as the words 'change to the system' meant a nightmare to the vendor. The vendor always tried to avoid changes all the time with justification: 'the system in the current form allows you to enrol the population and produce ID cards for them'. The client concern was that the software in its current version opens doors for many mistakes from user entry errors to incomplete business functions that were not recorded during the phase of required specifications. It is also worth mentioning that changes to the system when agreed were taking so long to implement. For example, adding a field to the system (job title) took one month to implement with a very high cost.
Stability	Can the software continue functioning if changes are made? The risk of unexpected effects of modifications.	As mentioned above, the system's complex architecture implied that a change in one place would affect many parts of the system. A change in one part of the system, would normally cause unexpected effects as a result of the modification.
Testability	Can the software be tested easily? The effort needed for validating the modified software.	In general, system (business) processes and functions were tested against specifications. However, from a technical perspective, the complex

		architecture of the system made it impossible to test many areas of the software. The vendor was pushing for the system to be accepted from a functional perspective (including the network set-up).
Portability	A set of attributes that bear on the ability of software to be transferred from one environment to another.	
Adaptability	Can the software be moved to other environments? The software's opportunity for adaptation to different environments (e.g. other hardware/OS platforms).	The software was designed and coded to operate within a unique environment of databases, operating systems and hardware. Most of the hardware used proprietary APIs' (Programming Applications Interface) to interface with the system. This automatically locked the system to only use the specified set of hardware.
Installability	Can the software be installed easily? The effort needed to install the software in a specified environment.	Though installation files and guides were available, the software architecture was not clear at all. All attempts made by the technical members failed in this regard. Despite several requests, the vendor felt that the system should not be installed by anyone other than the vendor himself.
Co-existence	Does the software comply with portability standards? Conformance is the degree to which the software adheres to standards or conventions related to portability.	The system did not comply with any portability standards other than the vendor's own.

(continued)

Replaceability	Does the software easily replace other software? The opportunity and effort of using the software in the place of specified older software.	The PRIDC software was expected to take over the current population register database maintained as part of the immigration system in the Ministry of Interior. However, this was a long-term objective. The software needed to go under several revolutions, before it could achieve this objective.

5. Reflection

'Some problems are so complex that you have to be highly intelligent and well-informed just to be undecided about them.'

Laurence J. Peter

Many researchers and practitioners argue that measurement is an essential issue in project and process management and an improvement on the logic that it is not possible to control what is not understood, and it is not possible to scientifically understand what is not measured (4). Using measurement practices may well increase the rate of project success to a higher level (30).

In the real world, however, this may be argued to be a valid issue at the organisation level, through not at the individual project level. That is to say that projects usually have very short term strategies with very tight deadlines and tend to be the result of opportunistic behaviour; where applying such measurement strategies may not be seen to add value, bearing in mind the time and cost associated with such measurement analysis activities. As the UAE national ID system will become the most critical system in the country as the main central hub for population identity cross-checking and service eligibility (i.e., online with a 24/7 availability requirement), it becomes important that the software undergoes a thorough quality check.

Taking into consideration the CBS nature of the system, some components were viewed to be more critical to undergo thorough quality checks, as a failure in different software components may lead to everything from public frustration to complete chaos when the card becomes 'the means' of accessing services. The evaluation study carried out here attempted to provide a short but thorough overview of the PRIDC system, and to measure the system's quality against ISO 9126 standard. Many limitations had been encountered that it would greatly help the project team to address before the final acceptance of the system from the vendor.

From the evaluation the system was found to have been developed as a component-based software system, but most importantly was observed to be a closed system. This closed architecture – although it was promised to work as prescribed in the specification documents – was viewed to likely cause the following major drawbacks in the short and long run:

1. The system supported only a few hardware vendors, as this was seen to result in the system losing a certain amount of autonomy and promoting it to acquire additional dependencies when integrating COTS components.

2. System evolution was not a simple 'plug in and play' approach. Replacing one component was more typically to have knock-on effects throughout the system, especially where many of the components in the system were black box components.

3. The system architecture forced the client to return again and again to the original vendor for additional functionality or capacity.

The closed architecture with the different proprietary platforms it incorporated were altogether more likely to slow down the pace of organisational business and process excellence as changes to the system would be expensive and extremely difficult to maintain.

The literature has not been kind to closed system architectures as research shows that such systems have proved to be too slow and too expensive to meet the rapidly changing market needs, as they restrict the level of quality that can be achieved (30, 31, 32, 33). However, some vendors and service providers strongly advocate standardised systems via closed architectures. Their argument is that these architectures are necessary in their systems' standards efforts and that the openness of the component-based approach leads to a chaos of choices and integration problems, and that they need to address the 'security' needs.

Moreover, over the long-term life of a system, additional challenges may well arise, including inserting COTS components that correspond to a new functionality and 'consolidation engineering' wherein several components may be replaced by one 'integrated' component. Below are further reflections on the major ISO 9126 quality attributes.

5.1 Functionality

The functionality factors were mainly checked against the system specification documents. However, it was discovered on the release of the first version of the software that many business functions were not covered in the specifications, resulting in the need for subsequent releases to address and fill the operational gaps. However, the evaluated software version in this study was not in an acceptable state, as it required additional enhancements to cover some of the additional business functions and rectify identified deficiencies, errors and bugs.

It is also worth mentioning that the overemphasis of security requirements during the specification phase contributed exponentially to the existing high complexity of the overall system, and its interoperability with other sub-systems. The fundamental problem concerning software development is defined as trying to understand the customer's sometimes unspoken needs and requirements and translate these into a tangible software solution.

The literature shows that one of the principal causes of information system failure is when the designed system fails to record the business requirements or improve the organisational performance. Researchers argue that such failures were because many organisations tended to use rule-of-thumb and rely on previous experiences (35). The vendor adopted the waterfall system development approach when it came to user requirements analysis and system implementation.

The vendor was reluctant to make any modification to the developed system, and was presenting a high cost impact on each change, even if it was a change to modify labels of text fields on user screens. The common response of the vendor was that 'the system is developed according to the agreed specification, and any deviation from that is probably going to have a cost impact'.

This attitude of the vendor opened the doors to long discussion meetings and arguments in this area, and slowed down the progress of the project, as changes were shelved for long periods as some got buried and lost in extensive project documents and long meeting minutes. However, system functionality is a temporary matter that can be resolved once attended to. The most critical items that needed to be addressed along with the functionality concerns were the areas discussed next.

5.2 Reliability

Software reliability is the probability that a software system will not cause the failure of the system for a specified time under specified conditions. The different tests carried out during the acceptance of deliverables relied on systematic software testing strategies, techniques, and processes, and software inspection and reviewing specifications. However, during this study, it was found very useful to incorporate less systematic testing approaches to explore the ability of the system to perform under adverse conditions.

5.3 Usability

The software seemed to have many usability concerns when system users struggled to understand system processes and functions, as minimal user documentation was available that did not cover the areas users needed most.

Extensive training was required to educate the users on the system, and much effort was required from the registration centre supervisors to support the users. The system was required to go through a major review to evaluate its usability. It also needed to be enhanced to follow a standard GUI methodology overall.

5.4 Efficiency

System processes and functions were checked against the time indicated in the specifications from a functional perspective. Nonetheless, code review was not possible because the source code was not handed over to the client at the time of carrying out this evaluation. Overall, the technical team had concerns about the capability of the system to provide acceptable performance in terms of speed and resource usage.

5.5 Maintainability

The complex architecture of the system made the analysis and diagnoses of discovered system faults and their maintenance so difficult that problems were usually passed to the development team in another country for investigation and preparation of stop-gap repairs.

Besides, the complex architecture acted also as a huge barrier to making urgent changes to the system as it required long analysis to evaluate the impact on the different components of the software, associated with an unrealistic cost impact of implementing such changes claimed by the vendor.

5.6 Portability

The system had many proprietary API's to interface with the different components of the system, locking the system into using a specified set of hardware. Installation files and guides did not enable the reinstallation of the system.

Overall, the system was observed not to comply with any portability standards other than the vendor's own, which could be carried out only by the vendor himself. The vendor was asked to add APIs to the system to allow the plug-in of new components to the system of both data and hardware.

6. Conclusion

'You don't drown by falling in the water; you drown by staying there.'

Edwin Louis Cole

As widely quoted in the literature, the application of software metrics has proved to be an effective technique for improving the quality of software and the productivity of the development process, that is, the use of a software metrics program will provide assistance in assessing, monitoring and identifying improvement actions for achieving quality goals (see for example: (3, 4, 5, 6, 7, 8, 9, 12, 14, 29, 35, 36, 37). In this study, the author attempted to use the ISO 9126 quality model to evaluate the PRIDC system; mainly from a product quality angle. See also Figure 4.10.

The study presented in this article contributed to a great extent in spotting some of the system deficiencies that were addressed prior to the final acceptance and handover of the system. It was also the author's observation that the project team, with the workload and responsibilities put on them, seemed to be overloaded and to have a confused idea of

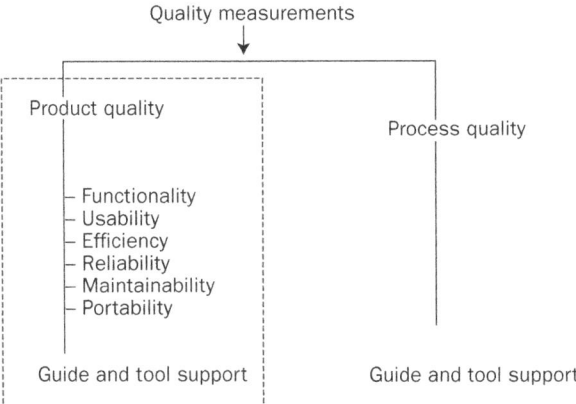

Quality measurements

Product quality

Process quality

– Functionality
– Usability
– Efficiency
– Reliability
– Maintainability
– Portability

Guide and tool support

Guide and tool support

Figure 4.10 Software quality metrics framework

how things should be done and achieved. Everybody wanted the project to conclude as quickly as possible, as everybody seemed also to be confident of the work produced by the vendor.

The use of the quality framework shown in this study can be a very useful and supportive methodological approach for going about software quality assessment. The ISO 9126 framework can act as a comprehensive analytical tool, as it can move beyond superficial evaluation to achieve a more thorough view of the system's strengths and weaknesses than can be provided by less systematic approaches.

When implementing big projects such as National ID schemes, project management and technical teams should use quality models for evaluating the overall architecture prior to the final acceptance of the system. As such, and if used as a guide in an early stage of the project, it can arguably provide a basis for informed and rational decision-making and have the potential to increase the project's success rate.

From a technical viewpoint, the ISO software quality metrics may also be extended throughout the phases of the software development lifecycle. The framework is designed to address the wide range of quality characteristics of the software products and processes enabling a better description of software quality aspects and its importance.

Acknowledgment

The author would like to thank Mr. Naorem Nilkumar for his contribution and technical input that improved the overall work presented in this article.

References

1. Al-Khouri, A. M. (2007) 'UAE National ID Programme Case Study'. *International Journal of Social Sciences* 1 (2): 62–69.
2. Folmer, E. and Bosch, J. (2006) 'A Pattern Framework for Software Quality Assessment and Trade-off analysis'. *International Journal of Software Engineering and Knowledge Engineering*. Available at: *http://www.eelke. com/research/literature/SQTRF.pdf*.
3. Bhatti, S. N. (2005) 'Why Quality? ISO 9126 Software Quality Metrics (Functionality) Support by UML'. *ACM SIGSOFT Software Engineering Notes* 30 (2).
4. Garrity, E. J. and Sanders, G. L. (1998) 'Introduction to Information Systems Success Measurement' (editors) in 'Information System Success Measurement'. Idea Group Publishing 1-11.
5. Grady, R. B. (1993) 'Practical results from measuring software quality'. *Communications of the ACM* 36 (11): 63–68.
6. Hastie, S. (2002) 'Software Quality: the missing X-Factor', Wellington, New Zealand: Software Education. Available at: *http://softed.com/Res- ources/ WhitePapers/SoftQual_XF-actor.aspx*.
7. Pfleeger, S. L. (2001) 'Software Engineering Theory & Practice'. Upper Saddle River, New Jersey: Prentice Hall.
8. Martin, R. A. and Shafer, L. H. (1996) 'Providing a Framework for Effective Software Quality Assessment – Making a Science of Risk Assessment'. 6th Annual International symposium of International council on Systems Engineering (INCOSE), Systems Engineering: Bedford, Massachusetts. Available at: *http://www.mitre.org/work/tech_transf-er/pdf/risk_assessment. pdf*.
9. Boehm, B. W., Brown, J. R., Kaspar, H., Lipow, M., MacLeod, G.J. and Merritt, M. J. (1973) 'Characteristics of Software Quality'. *TRW Software Series - TRW-SS- 73-09*, December.
10. McCall, J. A., Richards, P. K. and Walters, G. (1977) 'Factors in Software Quality'. Rome Air Development Center Reports 1, 2 and 3: NTIS AD/A-049 014, NTIS AD/A-049 015 and NTIS AD/A-049 016, US Department of Commerce.
11. Bertoa, M. F., Troya, J. M. and Vallecillo, A. (2006) 'Measuring the Usability of Software Components'. *Journal of Systems and Software* 79 (3): 427–439.

12. Valenti, S., Cucchiarelli, A. and Panti, M. (2002) 'Computer Based Assessment Systems Evaluation via the ISO9126 Quality Model'. *Journal of Information Technology Education*, 1 (3): 157–175.
13. Black, R. (2003) 'Quality Risk Analysis'. USA: Rex Black Consulting Services Available at: *http://www.rexblackconsulting.com/publications/Quality%20Risk%20Analysis1.pdf.*
14. Schulmeyer G. G. and Mcmanus, J. I. (1999) 'The Handbook of Software Quality Assurance' (3rd edition). Upper Saddle River, New Jersey: Prentice Hall.
15. Abran, A., Al-Qutaish, E., Rafa, J., Desharnais, M. and Habra, N. (2005) 'An Information Model for Software Quality Measurement with ISO Standards'. *SWEDC-REK, International Conference on Software Development*, Reykjavik, Iceland, University of Iceland 104–116.
16. Lau, K. K. (editor) (2004) 'Component-based Software Development: Case Studies'. *World Scientific (Series on Component-Based Software Development)*1.
17. Brooks, F. P. (1987) 'No Silver Bullet: Essence and Accidents of Software Engineering'. *Computer* 20 (4): 9-10.
18. Brown, A. W. (1996) 'Preface: Foundations for Component-Based Software Engineering'. *Component-Based Software Engineering: Selected Papers from the Software Engineering Institute*. Los Alamitos, CA: IEEE Computer Society Press, 7-10.
19. Brown, A. and Wallnau, K. (1996) 'Engineering of Component-Based Systems'. Proceedings of the Second International IEEE Conference on Engineering of Complex Computer Systems, Montreal, Canada.
20. Szyperski, c. (1997) 'Component Software: Beyond Object-Oriented Programming'. New York: Addison-Wesley.
21. Cai, X., Lyu, M. R. and Wong, K. (2000) 'Component-Based Software Engineering: Technologies, Development Frameworks and Quality Assurance Schemes'. *Proceedings APSEC 2000, Seventh Asia-Pacific Software Engineering Conference*, Singapore, December: 372–379. Available at: *http://www.cse.cuhk.edu.hk/lyu/paper_pdf/ap-sec.pdf.*
22. Brown, A. W. and Wallnau, K. C. (1998) The Current State of CBSE'. *IEEE Software* 155: 37– 46.
23. Kirtland, M. (1999) 'Designing Component-Based Applications'. Redmond, Washington: Microsoft Press.
24. Heineman, G. T. and Councill W. T. (editors) (2001) 'Component Based Software Engineering: Putting the Pieces Together'. Boston, MA: Addis on-Wesley.
25. Leavnesn, G. T. and Sitaraman, M. (2000) 'Foundations of Component-Based Systems'. New York: Cambridge University Press.
26. Richardson, R. 'Components Battling Component'. *Byte* 22 (11): 114.
27. Veryard, R. (2001) 'The Component-Based Business: Plug and Play'. London: Springer- Verla.
28. Pour, G. (1998) 'Component-Based Software Development Approach: New Opportunities and Challenges'. *Proceedings Technology of Object-Oriented Languages, TOOLS* 26: 375–383.

29. Godbole, N. S. (2004) 'Software Quality Assurance: Principles and Practice'. Oxford, UK: Alpha Science International.
30. Bass, L., Clements, P. and Kazman, R. (1998) 'Software Architecture in Practice'. Reading, MA: Addison Wesley.
31. Bosch, J. (2000) 'Design and use of Software Architectures: Adopting and evolving a product line approach'. Harlow: Pearson Education (Addison-Wesley and ACM Press).
32. Buschmann, F., Meunier, R., Rohnert, H., Sommerlad, P. and Stal, M. (1996) 'Pattern- Oriented Software Architecture: A System of Patterns'. New York: John Wiley and Son Inc.
33. Shaw, M. and Garlan, D. (1996) 'Software Architecture: Perspectives on an Emerging Discipline'. New Jersey: Prentice Hall.
34. Crosby, P. B. (1979) 'Quality Is Free: The Art of Making Quality Certain'. New York: McGraw-Hill.
35. Dromey, R. G. (1995) 'A model for software product quality'. *IEEE Transactions on Software Engineering* 21 (2): 146–162.
36. McCabe, J. T. (1976) 'A Complexity Measure'. IEEE *Transactions on Software Engineering* (2) (4): 308–320.
37. Möller K. and Paulish, D. J. (1993) 'Software Metrics'. London: Chapman and Hall Computing.

Electronic government in GCC countries: barriers and solutions

Ali M. Al-Khouri and Jay Bal

Abstract: This study investigated the practices of organisations in the Gulf Cooperation Council (GCC) countries with regards to G2C e-government maturity. It revealed that e-government G2C initiatives in the surveyed countries in particular, and arguably around the world in general, are progressing slowly because of the lack of a trusted and secure medium to authenticate the identities of online users. The authors concluded that national ID schemes will play a major role in helping governments reap the benefits of e-government if the three advanced technologies of smart cards, biometrics and public key infrastructure (PKI) are utilised to provide a reliable and trusted authentication medium for e-government services.

Key words: *E-Government, G2C, national ID, online authentication, biometrics, PKI, smart card.*

1. Introduction

Among the many promises of the Information Communication Technologies (ICT) revolution is its potential to modernise government organisations, strengthen their operations and make them more responsive to the needs of their citizens. Many countries have introduced e-government programmes that incorporate ICT and propose to transform several dimensions of their operations, to create more accessible, transparent, effective and accountable government.

Evaluating current practices, recent studies show that the implementation of e-government programmes is not a simple task as many, if not all, governments lack the fundamental infrastructure, organisational culture, understanding and resources for a transformation of the magnitude that e-governments require. Many researchers have addressed the technical and management issues surrounding e-government projects. Many others also have demonstrated the challenges associated with the implementation of e-government programmes, and put forward recommendations to overcome them.

Despite the variety of approaches that were proposed in the literature to handle government electronic services, not one proven solution or framework to build an e-government architecture appears to exist.

The objective of this study is to provide a short overview of the current literature in this research area and relate this information to the issues surrounding e-government initiatives. In principle, the study is designed to:

(1) Explore the potential applications of a national ID card and its suitability as a reliable medium to verify virtual online identities (if implemented with smart card, biometrics, and PKI technologies), and

(2) conduct a postal survey (followed by telephone interviews of executives) of organisations in the GCC* countries to understand their e-government practices and progress.

The finding of the study adds to the body of knowledge, as it draws a picture of the current practices, assesses the progress in the field of e-government and pinpoints the key obstacles and the degree to which national ID programmes can support the progress of G2C initiatives.

This paper is structured as follows. First a short overview of the current literature on e-government is provided to highlight current trends, patterns, and models for such initiatives, as well as the barriers to successful implementation. The following sections establish a link between national ID card schemes and e-government by looking at the technology requirements for enabling a reliable digital ID framework that can support and enable e-government development. Then the research survey methodology is explained, findings are presented and conclusions drawn.

---

* The surveyed organisations in this study were all from the southern Gulf countries; Bahrain, Kuwait, Oman, Qatar, Saudi Arabia, and the United Arab Emirates, often referred to as the Gulf Co-operation Council (GCC) countries.

2. The illusion of e-government

Citizens' experience with the 24/7 world of the private sector has fuelled demands for a similar experience with their governments; easy to deal with, available when you want them to be, a one-stop service that is personalised with simple completion of transactions online. This utopia bears little resemblance to most government's current capabilities; multiple agencies, multiple payment and delivery options, little coordination or standards, modest online functionality and variable customer service capabilities. Citizen demands are at odds with the current structure of most government agencies. Evidence is emerging, however, that when the government does go online successfully, patterns of interaction are dramatically changed.

In principle, the literature examines e-government activities in terms of the interactions between sectors of government, businesses and citizens (1). The matrix in Figure 5.1 shows the nine principal interactions. Some research studies also included employees in this spectrum. However, many researchers have considered the employee element to go under government activities.

| | Recipient of Services | | |
	Citizen	Government	Business
Citizen	Citizen to Citizen (C2C) e.g., small advertisement	Citizen to Government (C2G) e.g., tax declaration by single person or family	Citizen to Business (C2B) e.g., job exchange by job seekers
Government	Government to Citizen (G2C) e.g., benefit processing	Government to Government (G2G) e.g., transactions between Pas	Government to Business (G2B) e.g., procurement of PAs
Business	Business to Citizen (B2C) e.g., Online order in a shopping mall	Business to Government (B2G) e.g., tax declaration by private organisation	Business to Business (B2B) e.g., procurement through EDI

Suppliers of services (left axis label)

Figure 5.1 e-Government interactions

2.1 Government-to-government (G2G)

This represents the backbone of e-government. It involves sharing data and conducting electronic exchanges between governmental departments, rather than being focused on the specific agency or agencies responsible for administering programs and policies.

2.2 Government-to-business (G2B)

This includes both the sale of surplus government goods to the public, as well as the procurement of goods and services. It aims to work more effectively with the private sector because of the high enthusiasm of this sector and the potential for reducing costs through improved procurement practices and increased competition.

2.3 Government-to-citizen (G2C) – (the focus of this paper)

This provides opportunities for greater citizen access to, and interaction with, the government. This is what some observers perceive to be the primary goal of e-government. Thus, and from a G2C perspective, many government agencies in developed countries have taken progressive steps toward the web and ICT use, adding coherence to all local activities on the Internet, widening local access and skills, opening up interactive services for local debates, and increasing the participation of citizens in promotions and management of the territory (2).

Several approaches were proposed in the literature to handle electronic services. However, the literature provides not one proven solution or framework to build an e-government architecture. For this very same reason, e-government architecture development practices around the world vary according to several factors (the technical team experience, solution providers, consultants, budget, technological limitations, and so on), leaving those organisations with no choice but to go for a model and then enhance it based on new requirements and/or constraints.

To build these architectures, governments need to understand the complexity associated with the development and transition stages of e-government. One of the well-known models in the literature that outlines the stages of e-government development was developed by Layne and Lee (3) that outlines the stages of e-government development. In moving to the first two phases, government organisations are faced with technological challenges such as those in Figure 5.2. Stage three and four is where governments, instead of automation, transform their services and integrate processes and functions across the different levels of the government to create an integrated information base, implementing a 'one-stop-shopping' concept for its citizens.

The model assumes that e-government initiatives will require both horizontal and vertical integration; horizontal as e-government efforts must extend to all departments within a level of government (i.e., federal, state, local), and vertical as e-government initiatives must integrate across all levels of government. It was observed in the literature that some researchers used Layne and Lee's four phases and interpreted them as components of a maturity model to judge the maturity of the processes of an organisation and to identify the key practices that are required to increase the maturity of these processes – see for example (4). By focusing on a limited set of activities and working aggressively to achieve them, it is argued that the maturity model can steadily improve organisation-wide e-government processes and enable continuous and lasting gains in the e-government capabilities.

However, and according to various studies of e-government practices around the world, many researchers have found that such initiatives are stuck in phase one and two, far from the ideal integrated digital government (5). Researchers have identified many technical and organisational barriers challenging e-government progress to move up the ladder to stage three and four of Layne and Lee's model (see Figure 5.3).

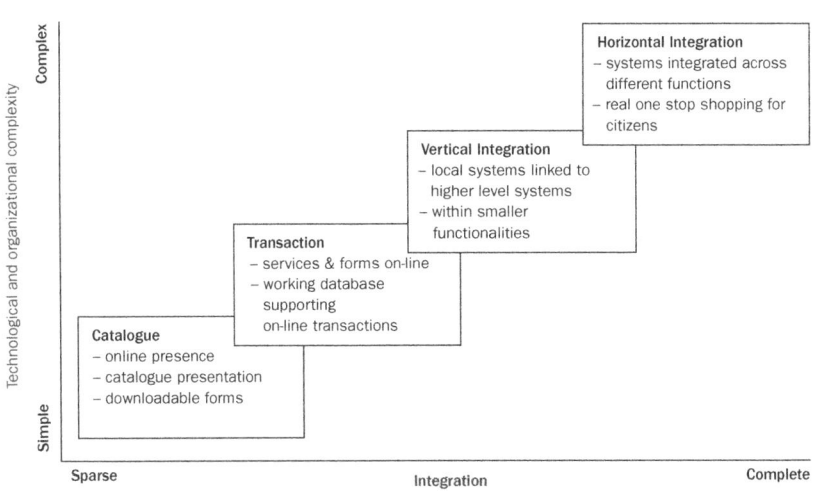

Figure 5.2   Dimensions and stages of e-government

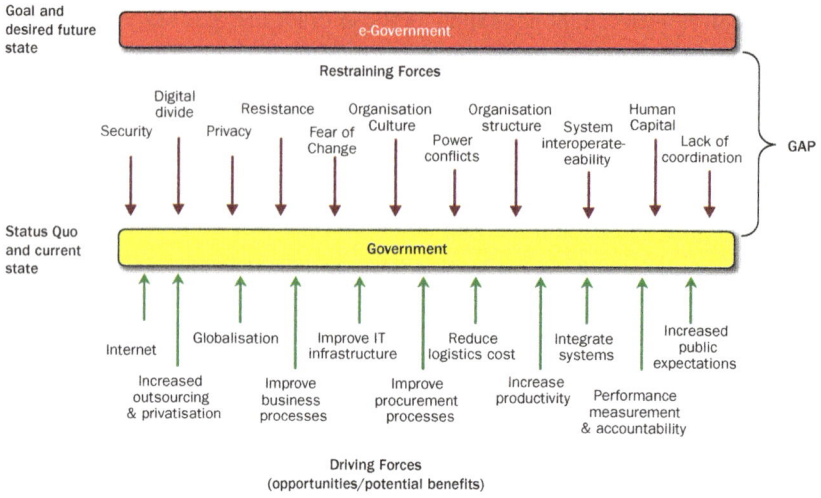

Figure 5.3 e-Government force analysis

It appears the current literature does not give enough attention to the need in e-government programs for identity verification, which is believed by the authors to be the foundation for G2C initiatives. It is argued by the authors that if governments are to complete phase two of the Layne and Lee model and enable a much larger and more comprehensive set of G2C transactions to take place online, they will need to ensure that citizens have the ability to authenticate themselves online and verify their identities – see also (6). Governments need to develop a clear vision of how they intend to authenticate individuals' digital identities (7). A digital identity is the representation of a human identity that is used in distributed network interaction with other machines or people (8).

Their vision needs to look at the different options available for building a digital identity management infrastructure that 'allows transactions in which the parties are separated in time and space while retaining the ability of these transactions to contain all of the human identity based attributes that transactions between people have always had' (8). Only with a robust digital identity infrastructure can the true power of G2C applications be released. Initiatives such as national ID projects are a key to G2C e-government progress, and a step towards building a secure digital infrastructure that can enable online identification and authentication.

The national ID project is seen by the authors as a good opportunity to build governments' central identity infrastructure component for e-government initiatives. The next section looks at how advanced technologies can support G2C e-government and provide a robust digital ID, as well as a solid foundation for developing secure applications and safeguarding electronic communications.

3. National ID and G2C e-government

National ID programs may well address many of the security issues related to electronic communications and the verification of online identities, provided that appropriate technologies are utilised.

This can also be realised by looking at one of the primary goals of such schemes, take for instance the UAE national ID project, which aims to improve the country's ability to accurately recognise peoples' identities through identification (1:N) and verification (also referred to as authentication) (1:1) methods as depicted in Figure 5.4 (9).

The key to G2C e-government is authentication, that is, the ability to positively identify and prove the authenticity of those with whom the government conducts business. Without authentication, other security

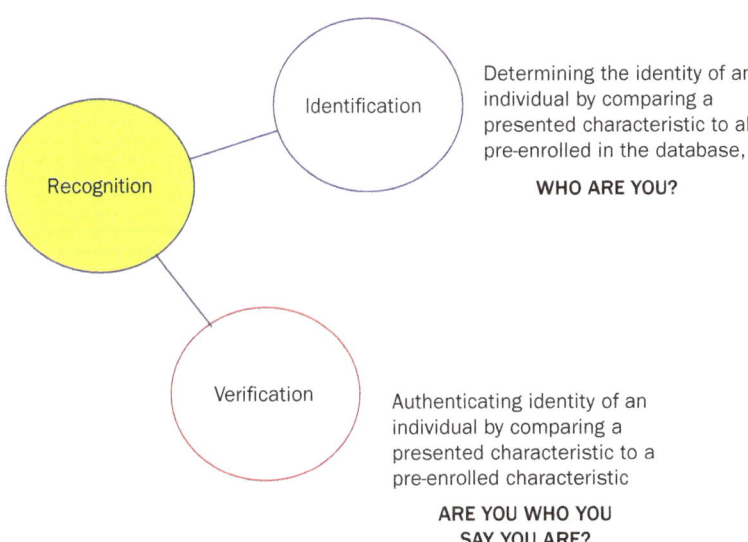

Figure 5.4   Identity recognition

measures put in place for many G2C transactions can be ineffective. To clarify this further, governments need varying levels of authentication strength based on the value or sensitivity of their online information or services, balanced against other considerations like usability, deployment, and budget. The discussion in this section is limited to the presentation of authentication levels that organisations may consider in their G2C initiatives and some of the advanced technologies that can make such requirements possible. It is important to heed that the essence of G2C e-government is that transactions occur between people that are represented by machines. The anonymity of these transactions makes it more difficult to identify the parties involved and to ensure a trusted business relationship.

Since all successful business relationships are based on trust, establishing online trust should be one of the primary goals of any e-government initiative. The focus must be on building a trust environment that provides a high level of data privacy, data integrity, and user authorisation. The real cornerstone of G2C e-business trust is authentication: that is, knowing with whom the government is doing business. PKI, smart cards, and biometrics (see Table 5.1) are the technologies that are believed to be the key components of the trust model to address both electronic transactions security and online identity authentication.

Combining these three technologies can provide the government with a three-factor authentication capability such as depicted in Figure 5.5:

Table 5.1   PKI, smart cards, biometrics

(1) Public Key Infrastructure (PKI):	State of the art in digital authentication and overall security infrastructure
(2) Smart Card:	A plastic card with an IC chip capable of storing and processing data that may also come with optional magnetic strips, bar codes, optical strips etc. viewed as an ideal medium for national ID schemes, e-government and e-commerce applications
Biometrics:	Allows individuals to meet their credentials and therefore enables the verification (authentication or identification) of people's identity using the unique properties of their physical characteristics

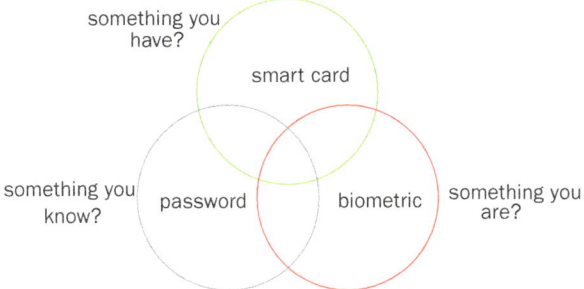

Figure 5.5   Three-factor authentication

1. A password to ascertain what one knows,

2. a token (smartcard) to ascertain what one has/possesses, and

3. biometric recognition (for example fingerprint or thumbprint) to ascertain who one biologically is.

As such, if passwords have been compromised, fraudsters need to get through another two levels of authentication to access a customers account. This would be difficult, if not totally impossible.

By requiring three forms of identification to access credentials, organisations will be able to bind users' digital identities to their physical identities which allows them to be more confident that the users are who they say they are, which should in turn give governments a high level of assurance of online identities.

The following three sections will introduce the three main technologies, namely: PKI, biometrics, and smart cards.

Due to the breadth and depth of the PKI subject, the discussion here is limited to addressing the online identity authentication issue. Several practical studies demonstrated that most of the e-government security requirements can be fulfilled through the public key infrastructure (PKI) security services.

PKI is defined as a system of computers, software and data that relies on certain sophisticated cryptographic techniques to secure on-line messages or transactions (10). The requirements imposing the need for additional security measures are either related to the hardware/ software infrastructure of the e-government platform (e.g. performance, availability, and so on), or to highly specialised-security critical applications (e.g. e-voting; anonymity, un-coercibility, and so on).

In principle, as depicted in Figure 5.6, PKI provides four key features to secure online transactions:

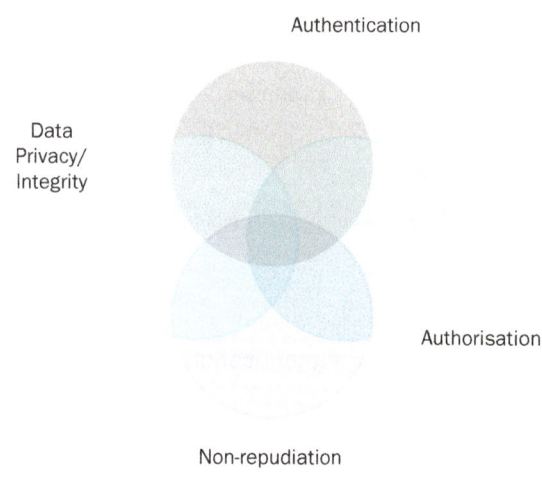

Authentication

Data Privacy/ Integrity

Authorisation

Non-repudiation

Figure 5.6   **PKI security framework**

- Authentication – to verify the user identity prior to an online exchange, transaction, or allowing access to resources (e.g., digital certificate*, public key certificate, biometrics, and so on).

- Data privacy/integrity – to ensure the confidentiality of information and that data is not altered as it moves around the public Internet (e.g. encryption**).

- Non-repudiation – to prove that an individual has participated in a transaction (e.g., digital signature). Only a two-factor authentication definitively binds a user's physical identity to his digital identity.

---

*A digital signature is sometimes referred to as an electronic signature, but is more accurately described as an electronic signature that is authenticated through a system of encryption logarithms and electronic public and private keys. A digital signature is often described as an envelope into which an electronic signature can be inserted. Once the recipient opens the document, the digital signature becomes separated from the document and the document can be modified. Thus, a digital signature only preserves the integrity of a document until it is opened.

**Encryption is a security method that transforms information into random streams of bits to create a secret code. There is software-based encryption such as Secure Sockets Layer (SSL) or Public Key Infrastructure (PKI). Hardware-based encryption, such as smart cards, is another type of encryption.

- Authorisation – to verify that the user has permission to participate in an activity, a transaction, or is allowed access to resources (e.g., cross reference public key certificate with a privilege through the use of policy management servers).

PKI provides a mechanism for binding cryptographic keys, used to encrypt and digitally sign messages to other credentials such as name, age or place of birth from key certificates and transporting those certificates around the Internet electronically.

A government agency can, for example, send messages using a citizen's digital certificate encoded with relevant public key that only that specific citizen can open.

3.1 Biometrics

Biometric technology can be used in identity management systems to identify someone in a population (known as 1:N matching) or to verify someone against his/her own details (known as 1:1 matching). Apart from being non-transferable among individuals, biometrics do not provide data about the person, but rather, information about the person.

When biometrics such as fingerprints or iris recognition are deployed in these contexts, for unique identification and for strong authentication, they provide an effective means for binding people to their identities.

In the context of a national ID scheme, the biometrics process allows a technique of padlocking the citizen to the card. In doing so, the card cannot easily be transferred to another individual. In particular, given the current focus on the use of biometrics in national identity cards, it sets out an architecture for strongly-authenticated identity cards that deliver (perhaps counter-intuitively) both enhanced security and enhanced privacy.

3.2 Smart Cards

In a smart card-secure environment, users are not locked into one form of authentication, such as the ever vulnerable password. Smart cards provide a mechanism for binding cryptographic keys to individuals, with appropriate authentication, so that when a key is used then the organisation or the individual can be certain of the identity of the person at the other end of the transaction or communication.

Mapping this to the earlier example of the government agency, when a person gets a message, he or she can put their smart card into their PC and punch in their PIN that will in turn let the smart card use the relevant key to decode the message. Depending on the configuration, if a user loses the smart card, it is inoperable without the biometrics.

Forged fingerprints can be weeded out with the use of the PIN. Smart cards allow on-card or off-card biometric matching. Off-card matching means that biometric authentication happens online where the biometric features are compared with backend databases.

On-card matching technology means that biometric features are compared with a stored template within the card. The template is stored exclusively in the secure smart card environment, which reliably protects sensitive personal data against unauthorised access.

On-card matching is an outstanding way of user authentication within security applications that meet the three paramount requirements of

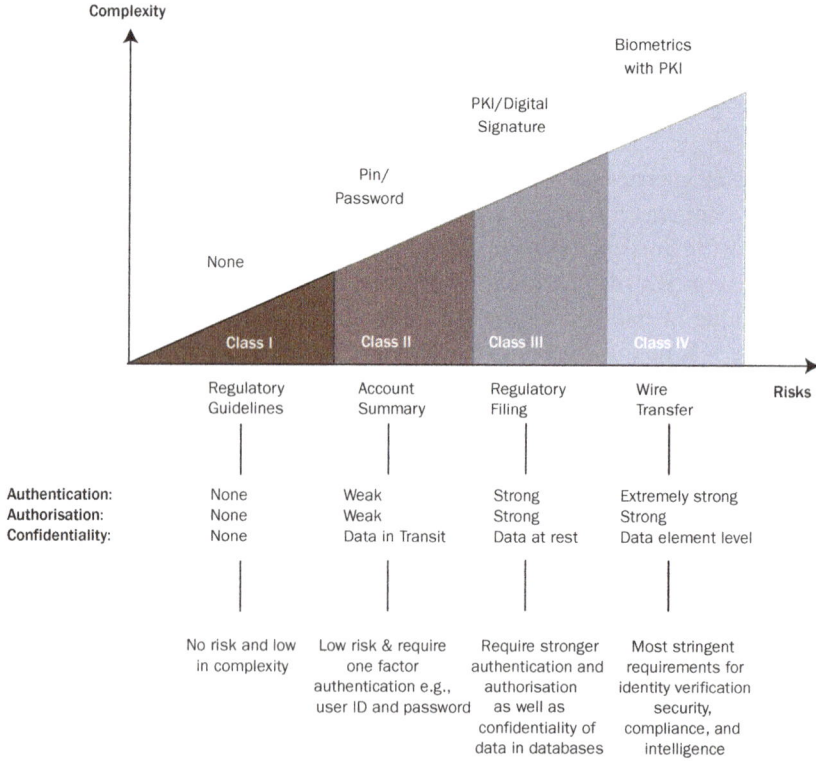

Figure 5.7 An example of types of authentication for G2C e-gov services

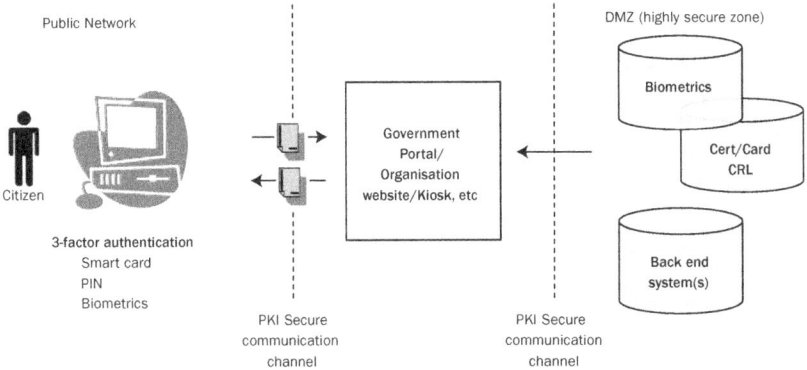

Figure 5.8 Conceptual model for electronic authentication

security, ease of use, and data privacy. Using the power of these three technologies, government organisations and businesses alike can use varying levels of authentication depending on the level of security required for a particular transaction (Figure 5.7).

Citizens with simple readers attached to their PCs (at home, work) or even kiosk machines can log on to the government Internet portal and perform various transactions online in off-card or on-card authentication modes (Figure 5.8).

The next sections present the research methodology and the findings from a survey conducted to assess the current status of e-government projects in the Middle East.

4. Research methodology

The data of this study was gathered by two principal methods: personal interviews and a questionnaire survey. A six-page questionnaire was designed, consisting of structured and semi-structured questions, to gather information and understand the surveyed organisations' practices in the field of e-government.

The questionnaire was pilot tested through telephone interviews with four senior executives and two managers in two organisations – following the recommendation of (11, 12, 13). These interviewees provided detailed feedback on the clarity of the questions and overall comprehensibility of the instrument.

The result of this pilot study led to some adjustments to the content and format of the questionnaire and terminology used in the survey. The updated questionnaire was then pre-tested on six interviewees in four organisations before being administered to all participating organisations (excluding those who took part in the pilot study).

A total number of 198 questionnaires (both in English and Arabic-language versions) were sent to the contacted organisations through the chief information officer or the IT/IS manager, as advised by the sites. The majority of the questionnaires were mailed to respondents with pre-paid envelopes, while the rest were either faxed or e-mailed to respondents. Sixteen questionnaires could not be delivered to the intended recipients and were returned by the postal service.

A total number of 60 questionnaire forms were returned (by postal mail and electronically), giving a response rate of 30.3 per cent. After evaluating the responses, it was found that eight responses were unusable owing to insufficient data. The removal of these unusable responses gave a total of 64 usable questionnaires (that is including 12 responses produced from the first pilot mailing), which represent an overall response rate of 30.5 per cent.*

In the sample of 64 participants, 26 government organisations were represented. The characteristics of the respondents are summarised in Table 5.2. The country with the most participants was the United Arab

Table 5.2 Responses by industry

Industry (Government)	No. of organisations
Oil/Petroleum	4
Medicine/Health	3
Transportation	4
Telecommunication	3
Finance/Insurance	4
Other	8

* The results of the pilot and final questionnaires were merged here since the changes made to the initial questionnaire were only to clear out ambiguity and change the arrangement of questions. The additional questions included in the final questionnaire were questions 27 and 28 (See Appendix: Research Questionnaire).

Emirates with a response rate of 38.1 per cent, followed by Bahrain with 32.5 per cent. Table 5.3 gives response characteristics for each site.

4.1 Telephone interviews

Respondents to questionnaires were asked if they could be contacted to provide some clarification and to be asked some additional questions for the purpose of improving the quality of the research information. It was also made clear that their right to anonymity would not be affected in either case. Out of the 64 respondents, 21 agreed to be interviewed, 18 of whom were executives and department directors, and five senior managers.

The initial draft of the questionnaire served as an interview guide to ensure that all the relevant questions were asked. Semi-structured, telephone interviews (transcribed for subsequent analysis) were administered to 19 individuals[*] in 12 organisations.

The semi-structured interviewing approach was developed to ensure that the research questions were properly addressed while allowing for 'probing' questions to gain even greater understanding and insight into the issues. The qualitative data obtained through telephone interviews helped to fine-tune the focus of the questionnaire survey and interpret its quantitative results.

Some additional follow-up interviews were also conducted by telephone and e-mail. All the interviewees were very friendly and were willing to share their experiences and ideas. Most of the interviews lasted between 20 and 30 minutes.

4.2 Measurement of variables

The questionnaire was divided into two sections to help break the monotony, ease problems of comparison and, most importantly, enable the arrangement of the questions thematically (11).

---

[*] Out of the twenty-one respondents who agreed to be interviewed, two senior managers declined later without any explanation.

Table 5.3 Survey response characteristics by region

Site (Distributed/ Returned/per cent)	Functional areas of returned surveys (self-reported)		Hierarchical levels of returned surveys (self-reported)	
Bahrain (34 / 11 / 32.4 per cent)	Corporate Mgmt	3	Executive	6
	IS/IT	3	Senior/mid mgr.	3
	Human Resources	1	Missing IDs	2
	Planning and Development	1		
	Finance	1		
	Missing IDs	2		
	Total	11	Total	11
Kuwait (29 / 8 / 27.6 per cent)	Corporate Mgmt	2	Executive	3
	IS/IT	2	Senior/mid mgr.	4
	Human Resources	1	Missing IDs	1
	Planning and Development	1		
	Finance	1		
	Missing IDs	1		
	Total	8	Total	8
Oman (31 / 9 / 29.0 per cent)	Corporate Mgmt	3	Executive	4
	IS/IT	2	Senior/mid mgr.	5
	Human Resources	1		
	Planning and Development	1		
	Finance	1		
	Missing IDs	1		
	Total	9	Total	9
Qatar (36 / 9 / 25.0 per cent)	Corporate Mgmt	2	Executive	4
	IS/IT	2	Senior/mid mgr.	3
	Human Resources	2	Missing IDs	2
	Planning and Development	1		
	Finance	1		
	Missing IDs	1		
	Total	9	Total	9

Table 5.3   Survey response characteristics by region (*cont'd*)

Saudi Arabia (38 / 11 / 28.9 per cent)	Corporate Mgmt	3	Executive	4
	IS/IT	1	Senior/mid mgr.	6
	Human Resources	1	Missing IDs	1
	Planning and Development	3		
	Finance	2		
	Missing IDs	1		
	Total	11	Total	11
United Arab Emirates (42 / 16 / 38.1 per cent)	Corporate Mgmt	6	Executive	9
	IS/IT	1	Senior/mid mgr.	6
	Human Resources	2	Missing IDs	1
	Planning and Development	3		
	Finance	1		
	Missing IDs	3		
	Total	16	Total	16
(210 / 64 / 30.5 per cent)	Total	64	Total	64

Part 1: This part was designed to be filled by the IT/IS department managers or the responsible department foreseeing the management and implementation of technology related services. The objective of this part of the questionnaire was to gain an understanding of some basic information about the IT infrastructure and technologies utilised to support the electronic strategies.

Part 2: This part of the questionnaire was designed for all respondents. The objective of this part was to understand the perceptions of both executives and other senior managers about e-government opportunities, obstacles and future plans, as well as their level of awareness.

5. Research findings

The following table summarises the research findings.

Table 5.4 Summary of research findings

1. How important do organisations perceive online presence and e-government initiatives?
Out of the 26 surveyed organisations, 20 had websites, four were planning to develop one, two indicated that they 'have no intention of developing one,' with the justification that they had no interaction with citizens. Overall, all respondents perceived e-government as a concept that gives them an opportunity to revolutionise their organisations.
2. A major driver for e-government projects?
Customer expectations and internal efficiency/cost reduction were found to be the most common drivers for e-government projects.
3. What plans/strategies do organisations have to go about e-government projects?
None of the responding organisations indicated having an e-government strategy, but rather a set of guidelines and short-term plans. These plans focused on augmenting internal operations, where G2C was left to the operational departments to implement. In most cases, IT departments were tasked to champion such projects.
4. Impact of e-government on organisations' operations.
e-Government was viewed to enable the government to appear as one unified organisation and provide seamless online services.
5. What is the greatest obstacle to e-government initiatives as viewed by organisations?
Security was found to be a major concern. The ability to verify online identities was seen to be the biggest obstacle when it came to G2C transactions.
6. Can national ID projects support e-government projects?
Many have viewed national ID projects to be mainly addressing homeland security issues and to replace existing cards such as driving licence, health card, bank cards, and so on. However, a common view was that their governments, whether through for national ID or other programmes, must put a solution into place to address the need for online authentication of individuals to support e-government progress.

5.1. How important do organisations perceive online presence and e-government initiatives?

The results of the survey revealed that almost 77 per cent of government organisations that responded had a website (See Figure 5.9). Out of the 23 per cent of the respondents who responded 'no' to having a website, more than 66 per cent planned to create one this year or early next year.

Out of the two organisations that did not have websites, one executive claimed that they did not see the need to having one because of the nature of the services of their organisations which requires the physical presence of customers. The other executive claimed to 'have no intention of developing one, because of online security concerns'.

On the other hand, around 57 per cent of those who responded as having a web presence indicated they had automated online services such as payment of fees, bills and fines (see Figure 5.10). Only three (21.4 per cent) organisations indicated having integrated their systems for limited online

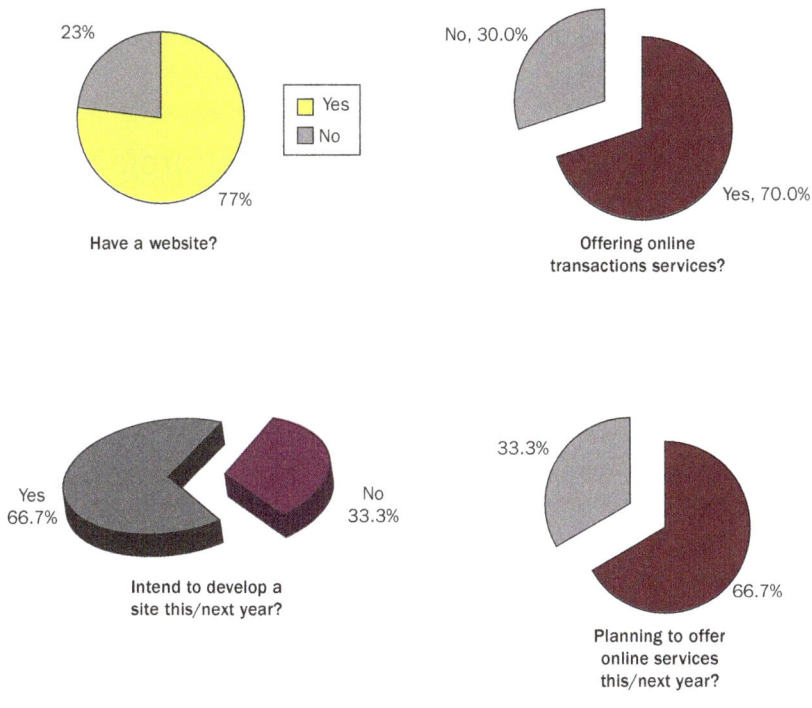

Figure 5.9   Online presence

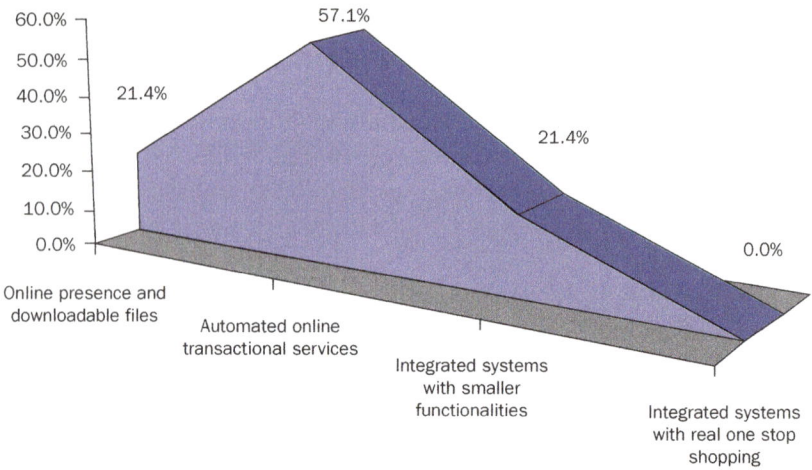

Figure 5.10 Organisations' own perception of their electronic operations

functions. This supports the findings of previous studies that most organisations are still in the cataloguing and transactional phases of the Layne and Lee model.

5.2 A major driver for e-government projects?

The majority of the respondents (85.9 per cent) indicated that the major driver for e-government projects was the (1) growing expectations of citizens for online services and (2) internal efficiency and cost reduction (see Figure 5.11).

Though not in the form of policies or legislation, responding organisations also reported that a significant amount of pressure is being applied by the government to better coordinate business processes and information flow among ministries and local departments.

Many of the interviewed executives said that because of the above two drivers for e-government, their organisations were in the process of planning to integrate automation, support collaborative business processes and streamline business operations.

Yet others expressed concerns over the ability to effectively integrate their systems and technically collaborate with other government organisations because of technical and security constraints.

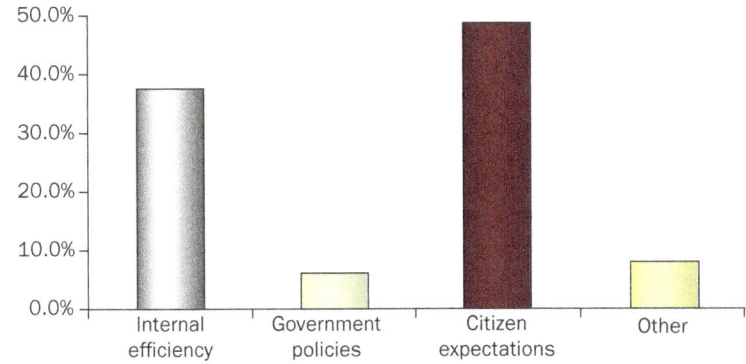

Figure 5.11   Main motives behind e-government projects

67 per cent of respondents agreed with the statement that 'online government services must be customer-centric'. The 21 per cent of respondents who answered 'no' to this statement indicated reasons such as confidentiality of records and current policies to hinder such concepts (see also Figure 5.12).

Many of them have also expressed their concerns about their inability to automate many of their services and put them online since identity verification was a prerequisite function, as one IT director explains, 'with the pressure we have from the top management to improve performance and offer online services, we are still struggling to address the online identity verification issue.'

As depicted in Figure 5.13, out of those organisations who indicated having a website, 26.7 per cent indicated that they had an IT infrastructure that could support limited online service plans. Not very surprisingly, 73.3 per cent responded 'no' to having one.

With follow-up phone calls to people in both categories, they indicated that they had invested in many advanced technologies to secure their services from any misuse, but the authentication of online identities was considered to be lacking certain elements in all organisations that made their infrastructure incomplete when it came to G2C transactions.

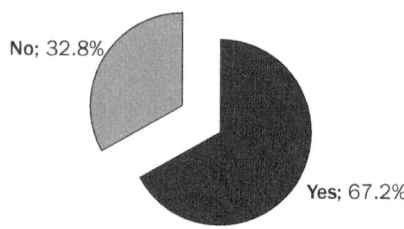

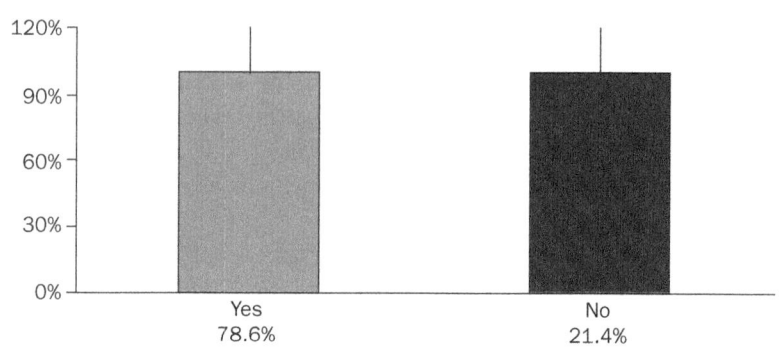

Figure 5.12 Re-engineering and customer-centric services

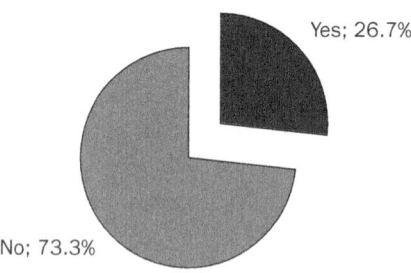

Figure 5.13 IT infrastructure readiness

5.3 What plans/strategies do organisations have to go about e-government projects?

Only 26.9 per cent of respondents indicated that they had an e-government plan, but no strategy, and that 61.9 per cent tasked information technology departments to create such plans and carry out their implementation (see also Figure 5.14).

Two of the organisations indicated having no clear vision or plan regarding their e-government, and said that they were in the process of appointing a consulting company to develop an e-government road map for their organisations.

During the course of interviews with the executive management, it was found that almost all organisations had a draft blueprint for going about e-government programs. However, they claimed that those strategies did not address the one-stop-shopping concept, but focussed more on internal

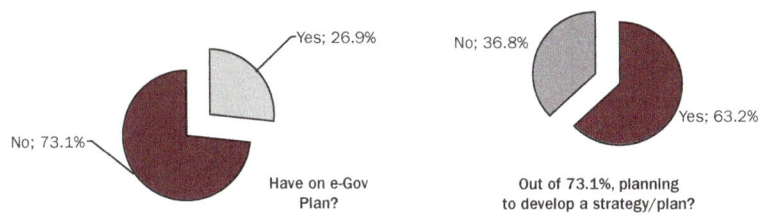

Yes; 26.9%
No; 73.1%
Have on e-Gov Plan?

No; 36.8%
Yes; 63.2%
Out of 73.1%, planning to develop a strategy/plan?

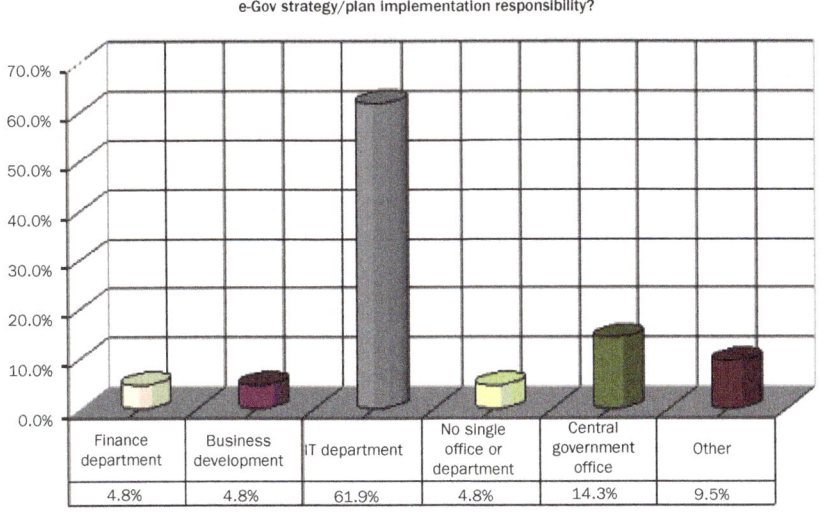

e-Gov strategy/plan implementation responsibility?

Finance department	Business development	IT department	No single office or department	Central government office	Other
4.8%	4.8%	61.9%	4.8%	14.3%	9.5%

Figure 5.14 e-Government strategy

organisational efficiency. Overall, many of the current e-government plans and strategies were believed not to address the G2C aspect, and were left to the ministries and other government departments to address.

Some organisations indicated that the information technology departments had a better understanding of what e-government involves, as one explains, 'the IT department knows more about these technological projects. We tasked them to coordinate with other departments to create e-government plans. In this way, the other departments can focus on their core business'. (Translated from an interview). This was a common and shared view among many of the interviewed executives.

This is also cited in literature as one of the key reasons that system projects fail. Information technology people tend not to know much about business goals and strategy. Hence organisations get IT systems that are not aligned with business strategy, which is the most common cause for project failure.

5.4 Impact of e-government on organisations' operations

More than 30 per cent of organisations indicated that e-government projects have increased the demand for forward thinking management and technical staff. It was also cited that such demand and skill shortages in different management and technical fields put upward pressure on wages. Follow-up phone calls were made to get some clarification from those who reported a reduction in the number of staff as a result of e-government programs.

The feedback received was that some of the re-engineering activities automated many of the internal processes, thus reducing the size of the work force. However, they also indicated that e-government projects had placed greater emphasis and demands on the need for solid project management and business process analysis skills, as well as the technical staff who could manage and administer complex technical systems.

It was also indicated during the interviews that many of the e-government projects caused a sharp rise in the use of outside contractors and consulting companies due to the complexity of the projects and to meet pressing deadlines. There was also this common view among many of the participants that e-government could improve the traditional service channel strategies, which enabled governments to appear as one unified organisation and provide seamless online services. The following factors were also among the cited impacts of e-government as recorded during the course of the interviews (see also Figure 5.15):

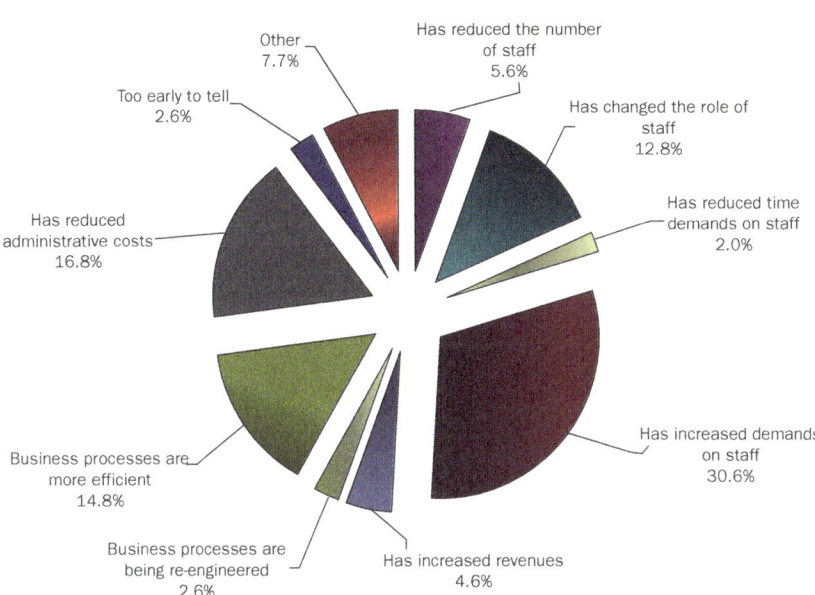

Impact of e-Gov on organisations?

Other
7.7%

Too early to tell
2.6%

Has reduced the number
of staff
5.6%

Has changed the role of
staff
12.8%

Has reduced
administrative costs
16.8%

Has reduced time
demands on staff
2.0%

Has increased demands
on staff
30.6%

Business processes are
more efficient
14.8%

Business processes are
being re-engineered
2.6%

Has increased revenues
4.6%

Figure 5.15 e-Government impact on organisations

- Improving business productivity (simplification of processes).
- Efficiency and improvements in processing internal activities, as well as public administration operations.
- Reducing expenditure through savings on data collection and transmission.
- Sharing of data within and between governments.
- Promoting an information society.
- Public management modernisation and reform.
- Enabling citizen engagement.
- Promoting open and accountable government.
- Preventing corruption, as this promotes more transparency.

5.5 What is the greatest obstacle to e-government initiatives as viewed by organisations?

Quite surprisingly, not one organisation indicated public concern over their privacy to be an obstacle to e-government projects. This may be due to the cultural and demographic nature of the countries in question. However, most organisations indicated to be using secure socket layer (SSL)* capabilities to ensure the privacy of information, especially for financial transactions and the transmitting of sensitive information.

As depicted in Figure 5.16, the majority of respondents (60 per cent) considered security issues to be the primary obstacle to their e-government projects, whereas 24 per cent indicated a lack of strategic direction and information about e-government applications to be the second most concerning and challenging issue. It was obvious that security was a common concern among the interviewed executive management in all surveyed organisations.

Although organisations indicated to be using many security technologies, online identity verification was stated as the biggest concern that led to

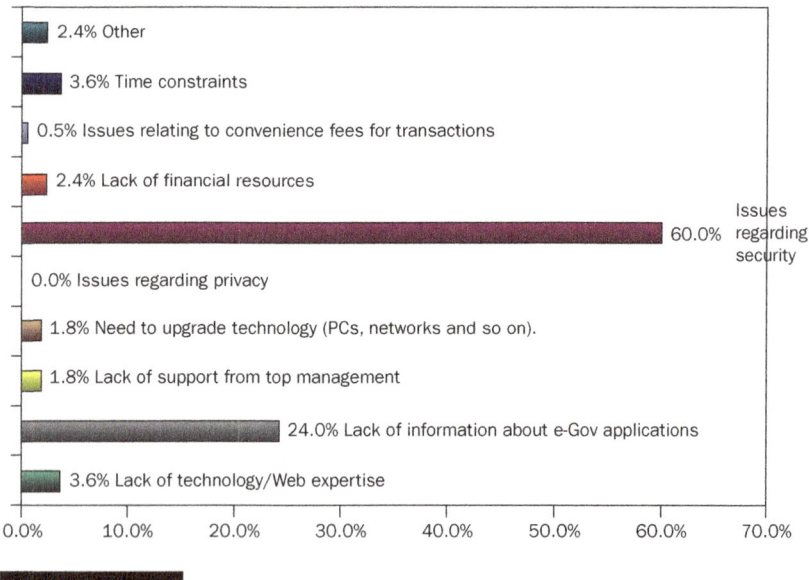

Figure 5.16   E-government obstacles

---

* With SSL, data is encrypted or scrambled prior to sending it and decrypting it on the receiving end. By encrypting the data as it travels through the Internet, it is virtually impossible for the transaction to be translated if intercepted.

slowing down their 'e-services plans' where identity assurance was required. This finding is consistent with the findings of Javelin Strategy and Research's 2005 Identity Fraud Survey Report published in 2005.[*]

Other organisations have indicated that although some services require authentication of their identities, and because of pressure from their top management, they are offering these services online, but using intermediaries such as postal or courier services to authenticate peoples' identities before delivering government documents/products to them.

As a part of the study, the online identity problem was further investigated. The interviews showed that many of the surveyed organisations faced transactions where people presented false credentials or those belonging to others to take advantage of some of the services the government provides. Three organisations have indicated that they have pulled back some of the online services they provided on the Internet after discovering that some people provided false credentials to gain access to sensitive information and benefit from some of the government online services. This area was noted as a common concern at some sites, as one executive said:

'Though we have invested a great deal in information technology and communication security, we are being challenged with attempts from some people trying to play around and take illegal advantage of the services we offer on the Internet'.

Another interviewee said:

We definitely need an identity management solution that guarantees to us the identity of those interacting with us online. Putting legislation in place that criminalises identity theft activities could be one part of the solution. But bear in mind that all those who perform such activities know that they are breaking the law. We need a mechanism to authenticate those people online.

More than 50 per cent of the respondents indicated to be utilising personal details or passwords to authenticate online individuals (see also Figure 5.17). None of the respondents indicated to be using PKI or biometric technologies for their online services. However, many of the IT department executives indicated that they are currently studying the possibility of introducing PKI and smart card technologies to address this growing area of concern.

---

[*] Published in January 2005, this report was co-released by Javelin Strategy and Research and the Better Business Bureau, and served as an update to the Federal Trade Commission's (FTC) 2003 Identity Theft Survey Report.

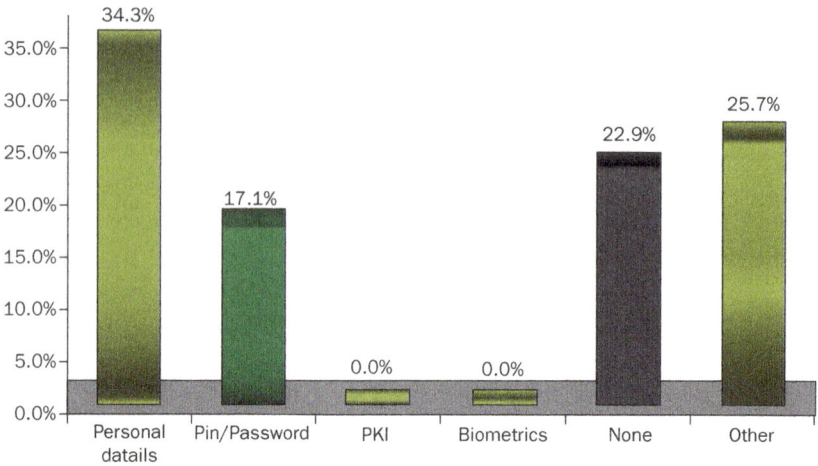

Figure 5.17 Utilised technologies for virtual identity authentication

5.6 How can national ID projects support e-government projects?

Though with some variations in their level of confidence, more than 55 per cent of respondents seemed to have confidence in biometrics to address the need for online verification. Around 38 per cent had some doubts about its suitability for online usage, whereas less than seven per cent indicated to having no confidence at all (see also Figure 5.18). It was also found during the interviews that some organisations were using smart cards and biometrics for authentication applications, both for internal access control purposes, as well as for some public services such as airports'. *

---

* In Dubai International Airport in the UAE, the electronic gate (e-Gate) project was launched in 2002 to allow frequent flyers fast access through immigration via electronically controlled gates. This fully automated passport control system replaces manual checks with smart cards and fingerprint technology to identify and clear registered passengers. It is also the intention of the government to use the new national ID card, and the thumb prints stored in the chip of the smart card for automatic immigration clearance without the need for the registering for the e-Gate service anymore.

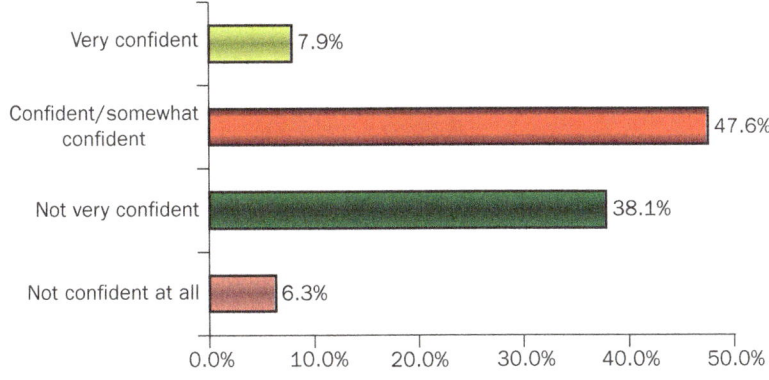

Figure 5.18   Level of confidence in biometrics

As depicted in Figure 5.19, more than sixty-four per cent of respondents viewed national ID projects as more likely to address homeland security than (online) identification of people.

Only 29.7 per cent indicated that they thought national ID projects would support e-government projects, and 51.6 per cent indicated their lack of knowledge in this regard.

Many of the interviewed executives believed that their governments needed to put in place appropriate mechanisms to identify people online, as one executive explains:

> I agree with the fact that the government has the responsibility to provide its population with identification means that proofs their identities and who they are. Today, there is pressure from the top to automate our operations and put the 'e' in our services. Whether through a national ID or other programmes, the government should provide the people with an 'e' identity that we can use to authenticate them online (translated from the interview).

A common view among the interviewees was also that the new smart ID card will allow the citizens and residents to authenticate themselves in an easy and completely secure electronic way whenever they access e-government applications. Another claimed advantage of the new card was that it will allow individuals to put their own electronic signature to digital documents such as declarations or application forms, which will have the same value and legal status as the documents that are nowadays signed by hand.

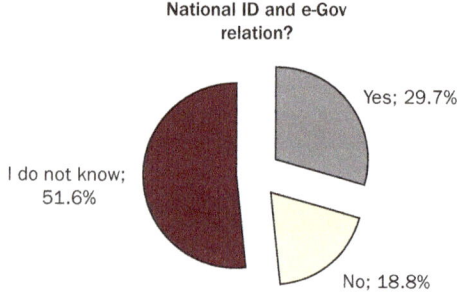

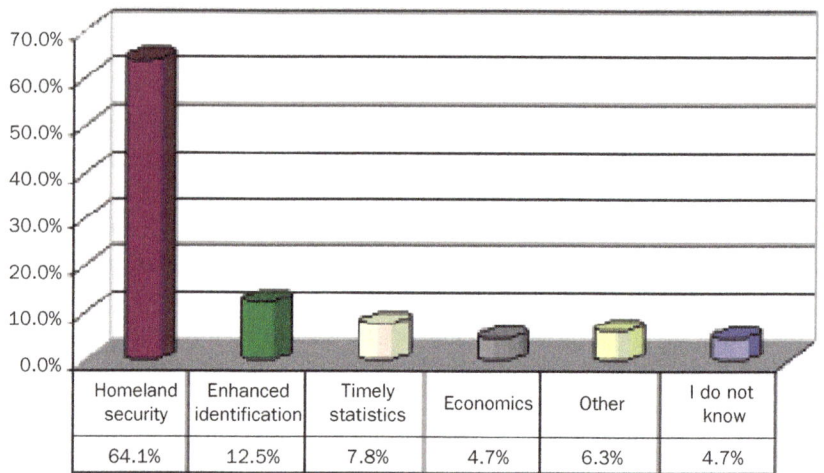

Understanding of National ID projects?

	Homeland security	Enhanced identification	Timely statistics	Economics	Other	I do not know
	64.1%	12.5%	7.8%	4.7%	6.3%	4.7%

Figure 5.19 Awareness of the relationship between national ID projects and e-government

6. Discussion and conclusion

The findings of this study are seen to be critical and have several implications for practitioners, especially if attempting to understand some practices related to G2C e-government in the GCC countries. Overall, the findings suggest that organisations need to have a more global view of what e-government is all about as many organisations tasked their IT departments to champion e-government projects. e-Government was seen as more of an automated activity.

This study shows that G2C e-government initiatives in the surveyed countries is progressing but in slow motion because of the lack of a trusted and secure medium to authenticate the identities of online users. In the surveyed organisations, many managers stated that the lack of a reliable authentication medium was preventing them from enabling many of their services to go online.

Consistent with the literature, the analysis of questionnaire responses and data collected from interviews with managers revealed that the use of passwords remains one of the most popular approaches used currently to address online authentication requirements.

With few variations in the perceptions of their impact, many of the obstacles to e-government presented earlier were highlighted by management in the surveyed organisations.

Security and overall system integration were by far the most widely quoted obstacles. Many of the interviewed managers indicated having computerised almost all their administrative functions, and in many cases their core business and support functions as well. Many organisations indicated that they had formed review committees to review their corporate plans and facilitate communication between departments, and to oversee the overall programme implementation. However, they appeared to have no structured approach to e-government strategy formulation and development.

Though each organisation had constituted a body in the form of a committee or department to carry out the 'e-readiness' assessment and thereafter draft a strategic plan for the implementation of e-government, it seemed, according to the interviewed executives to focus merely around G2G operations. The G2C was left to individual ministries and departments to implement.

6.1 The smart ID card and e-government

As viewed by many of the survey respondents and interviewees, governments must take the responsibility for putting in place a reliable identity management infrastructure. With the rapid evolution of technologies, governments need to introduce new and stronger means of identification and authentication for its population.

Traditional paper and conventional ID cards do not cope with the nature of the e-government environment, which requires advanced technologies to authenticate virtual identities over the web. Electronic authentication must be viewed as a fundamental part of the security infrastructure needed for the safe delivery of online government services, which gives both the user and the service provider confidence in the identity of the other party in a transaction. It is argued by the authors that initiatives such as national ID card schemes can very well address this requirement and can bring answers to many of the security concerns. The UAE national ID programme is a good example, as it aims to build a robust and secure national identity infrastructure.

The roll-out of this new national identity card in the UAE marks a major milestone in the development of e-government due to the nature of technologies it utilises biometrics, smart cards and public key infrastructure. The use of these technologies is seen to provide more secure and reliable G2C electronic authentication services. Indeed, such schemes pave the way for greater penetration and usage of government services and reap the promising benefits of e-government. It cannot be emphasised enough that for governments to fulfil their critical functions, they must be able to authenticate their citizens' claims about their own identities and characteristics (4). As digital government becomes a reality, the need for reliable digital identifiers becomes increasingly urgent.

6.2 Further research

This study was aimed only at organisations in the GCC countries. However, additional work must be carried out if a better understanding of worldwide e-government programmes is to be established. Some areas in which further research may yield valuable insights for more comprehensive understanding and assist management in determining optimal courses of action are:

1. A follow-up study in the same countries with a larger sample of organisations to gain insight into their perspectives and practices in the field of e-government development and implementation and to test the findings of this study.

2. A similar study should be conducted in other countries that could show the findings reported here are indeed universal and thereby might increase the robustness of the findings.

3. A study to shed light on the different views on power and control in organisations in relation to e-government adoption from both theoretical and empirical perspectives.

4. Field research for testing the impact of each identified obstacle on e-government programmes. This, in turn, should give a greater

understanding of those obstacles and pave the way to put forward appropriate, and/or develop, frameworks that can overcome such obstacles.

5. Understanding of the suitability of the national ID card with further detail as an authentication medium of online users.

As explained earlier, the authors intend to carry out a short practical study on the use of national ID cards (item five above) as a medium for online identity verification, in a separate study.

Acknowledgment

The authors would like to thank those who participated in the survey. They also would like to extend their gratitude to the reviewers who provided feedback that improved the overall structure and quality of this paper.

Appendix: research questionnaire

The purpose of this questionnaire is to investigate issues related to the field of e-government and its practices in GCC countries as part of a research project. This form is also available on the Internet for electronic submission. The website can be found at: *http://www.alkhouri.itgo.com/research/questionnaire.ht-ml*. If you have any queries about this questionnaire or need additional information, please contact the researcher at the following e-mail address: alkhouri@itgo.com.

Date: _____

Name: _____

Position: _____

Organisation: _____

Telephone: _____

Please note, you do not have to provide your telephone number if you prefer not to, and/or any of the other information.

<u>All information will remain strictly CONFIDENTIAL</u>

Part A: This section is to be completed by IS/IT managers.

1. Does your organisation have a website?

    Yes / No *(Delete as appropriate)*.

    If 'No', then do you intend to develop a website by this year or early next year? Otherwise, go to question 2.

    Yes / No *(Delete as appropriate)*.

2. Are you offering any online transactional services?

    Yes / No *(Delete as appropriate)*.

    If 'No', are you planning to offer any online service(s) by next year? Otherwise, go to question 3.

    Yes / No *(Delete as appropriate)*.

3. Has your organisation performed any re-engineering of the manual processes/services before offering it/them online?

    Yes / No *(Delete as appropriate)*.

4. How do you perceive your organisation's online services in the following categories?

a. Online presence and downloadable files	[ ]	*(Please tick as appropriate)*.
b. Automated online transactional services	[ ]	
c. Integrated systems with smaller functionalities	[ ]	
d. Integrated systems with real one stop shopping	[ ]	
Other		
(Please specify)		

5. What security methods do you use to secure online transactions?

    a. Personal details   [ ]   *(Please tick as appropriate).*

    b. Pin/Password   [ ]

    c. PKI   [ ]

    d. Biometrics   [ ]

    e. None   [ ]

    Other   _____

    (Please specify)   _____

6. Does your organisation have an IT infrastructure that supports your online services plans?

    Yes / No *(Delete as appropriate).*

    If '*no*', then please clarify, otherwise go to question 7.

Part B: This section is to be completed by all.

e-Government is the delivery of services and information to residents 24 hours a day, seven days a week.

7. *From your current work position, what do you think are the main motives behind e-government initiatives in your organisation?*

    a. Legislative requirements   [ ]   *(Please tick as appropriate).*

    b. Technological progress   [ ]

    c. Efficiency   [ ]

    d. Cost effectiveness   [ ]

    e. Services to citizens   [ ]

    f. Constituent/(citizen) demand   [ ]

g. Don't know   [ ]

Other   _____

(Please specify)   _____

8. *Do you agree that government services must be customer-centric?*

                                             Yes / No

Please specify _____

9. *Does your organisation have an overall e-government strategy and/or master plan to guide its future e-government initiatives?*

                Yes / No   *(Delete as appropriate).*

If '*no*', are you planning to develop a strategy/plan in the next year? Otherwise, go to question 9.

                Yes / No   *(Delete as appropriate).*

10. *Who has overall responsibility for implementing this strategy or plan or currently looking after e-government initiatives? (Tick only one)*

a. Finance department   [ ]   *(Please tick as appropriate).*

b. Business development   [ ]

c. IT/IS department   [ ]

d. No single office or department   [ ]

e. Central government office (specific Ministry or government department)   [ ]

Other   _____

(Please specify)   _____

11.   *How has e-government changed your local government?*
      (Tick all that apply).

      a. Has reduced the number   [ ]   *(Please tick as*
      of staff   *appropriate).*

      b. Has changed the role of   [ ]
      staff

      c. Has reduced time   [ ]
      demands on staff

      e. Has increased demands   [ ]
      on staff

      f. Has increased revenue   [ ]

      g. Business processes are   [ ]
      being re-engineered

      h. Business processes are   [ ]
      more efficient

      i Has reduced   [ ]
      administrative costs

      j. Too early to tell   [ ]

      Other   _____

      (Please
      specify)   _____

12.   *Please give an indication of your level of satisfaction with*
      the services provided by your own orgnaisation?

      Tick only one on a scale where 1 = very unsatisfied and 5 =
      very satisfied

      Very unsatisfied   1   2   3   4   5   very satisfied

13.   *Which, if any, of the following barriers/obstacles to*
      e-government initiatives has your local government
      encountered?
      Rate the following on a scale of 1 – 7.

      a. Lack of technology/   [ ]
      Internet expertise

      b. Lack of information   [ ]
      about e-government
      applications

c. Lack of support from top [ ]
management

d. Need to upgrade [ ]
technology (PCs, networks,
and so on)

e. Issues regarding privacy [ ]

f. Issues regarding security [ ]

g. Lack of financial [ ]
resources

h. Issues relating to [ ]
convenience fees for
transactions

i. Time constraints [ ]

Other _____

(Please
specify) _____

14. *In your opinion, what is the purpose of national ID projects?*

a. Homeland security [ ] *(Please tick as*
appropriate)

b. Enhanced identification [ ]
environment

c. Timely statistics [ ]

d. Economics [ ]

e. I don't know [ ]

Other _____

(Please
specify) _____

15. *If a biometric were used in these situations, how confident*
would you be that this technique would guarantee the identity
of online users?

a. Not confident at all [ ] *(Please tick as*
appropriate)

b. Not very confident [ ]

c. Somewhat confident [ ]

e. Very confident [ ]

Other _____

(Please specify) _____

16. *Do you think that if appropriate technologies such as PKI and biometrics were utilised, the national ID project would support your e-government initiatives by means of providing a safe and secure verification environment?*

a. Yes [ ] *(Please tick as appropriate)*

b. No [ ]

c. I don't know [ ]

17. *What would be the purposes of a national identity card?*

a. To prevent identity theft? [ ] *(Please tick as appropriate)*

b. For voting purposes? [ ]

c. To combat terrorism? [ ]

e. To facilitate international travel? [ ]

f. To replace many documents with a single card? [ ]

g. To access government services? [ ]

h. To combat illegal immigration? [ ]

Other _____

(Please specify) _____

18.   *If you have any additional comments which you feel would be helpful to this study, and in particular any difficulties, important factors or considerations which have not been mentioned, please state them here.*

_____

_____

If your answer is 'Yes', please make sure that you have included your telephone number on the first page.

We are very grateful for your help. Please return the completed questionnaire in a stamped addressed envelope to:

Ali M. Al Khouri, P.O.Box: 27126, Abu Dhabi, United Arab Emirates

Thank you once again for completing this questionnaire.

References

1.   Chadwick, A. and May, C. (2003) 'Interactions between states and citizens in the age of the internet: e-government in the United States, Britain and the European Union'. *Governance: an International Journal of Policy, Administration and Institutions* 16 (2): 271–300.
2.   Graham, S. and Aurigi, A. (1997) 'Virtual Cities, Social Polarisation, and the Crisis in Urban Public Space'. *Journal of Urban Technology* 4 (1): 19–52.
3.   Layne, K. and Lee, J. W. (2001) 'Developing Fully Functional E-Government: A Four Stage Model'. *Government Information Quarterly* 2: 122–36.
4.   Windley, P. J. (2002) 'eGovernment Maturity' USA: Windleys' Technolometria, Available at: *http://www.windley.co-m/docs/eGovernment%20Maturity.pdf*
5.   Devadoss, P.R., Pan, S. L. and Huang, J. C. (2002) 'Structurational analysis of e-government initiatives: a case study of SCO'. *Decision Support Systems* 34: 253–269.
6.   Leigh, A. and Atkinson, R. D. (2001) 'Breaking Down Bureaucratic Barriers: The next phase of digital government'. USA: Progressive Policy Institute.
7.   Windley, P. (2005) 'Digital Identity'. USA: O'Reilly Media, Inc.
8.   Digital ID World (2004) 'What is Digital Identity?' Available at: *http://www. digitalidworld.com/local.php?op=view&file=a- boutdid_detail*
9.   Al-Khouri, A. M. (2007) 'UAE National ID Programme Case Study'. *International Journal Of Social Sciences* 1 (6): 62–69.

10. Lambrinoudakis, C., Gritzalis, S., Dridi, F. and Pernul, G. (2003) 'Security requirements for e-government services: a methodological approach for developing a common PKI-based security policy'. *Computer Communications* 26: 1873–1883.

11. Hoinville, G. and Jowell, R. (1978) 'Survey Research Practice'. London: Heinemann Educational Books.

12. Oppenheim, A. N. (1992) 'Questionnaire Design, Interviewing and Attitude Measurement'. London: Pinter.

13. Shipman, M. (1988) 'Information through asking questions'. *The Limitations of Social Research* (third edition). London: Longman: 78–115.

14. Camp, L. J. (2003) 'Identity in Digital Government – A research report of the digital government civic scenario workshop' Cambridge, USA: Kennedy School of Government. Available at: *http://www.ljean.com/files/id-entity.pdf*

Digital identities and the promise of the technology trio: PKI, smart cards and biometrics

Ali M. Al-Khouri and Jay Bal

Abstract: This article looks at one of the evolving crimes of the digital age; identity theft. It argues and explains that if three key technologies were implemented together, namely biometrics, smart cards, and PKI, then they could deliver a robust and trusted identification and authentication infrastructure. The article concludes that this infrastructure may provide the foundation for e-government and e-commerce initiatives as it addresses the need for strong user authentication of virtual identities.

Keywords: *Identity theft, e-government, G2C, online authentication, biometrics, smart cards, public key infrastructure.*

1. Introduction

"My identity has been stolen – I am all over the internet!!!
It's awful being me....."

Identity theft has become the fastest growing crime in the world (1, 2). Undoubtedly, the expansion and increasing sophistication of identity theft threatens the adoption of many strategic information technology (IT) initiatives such as e-government and e-business (3, 4, 5). Identity theft is an activity that takes place when an individual's personal details are taken over or stolen by someone else in an attempt to impersonate him/her and to have access to particular information or services, to perform financial transactions, or even commit crimes. Identity theft has many and increasing links to organised crime.

As recently as ten years ago, people would research to find someone who had died before they ever had a job. They would then apply for a copy of the birth certificates in the names of those dead people, and use them it to obtain other ID documents. However, with the advances in the field of information technology, identity theft has become much easier. For instance, more than 30 websites offer fake ID's for sales from as little as $40 for a social security card, $79 for a birth certificate and $90 for a driving license from any US state.

Websites such as *www.fake-id.org* and *www.phonyid.com* offer driving licenses with similar security features issued by the US government from all 50 US states for $100 each, as well as Canadian ID. Several hundred dollars buys a complete ID set, including a military ID and a college diploma. Use of false identification is considered to be a significant threat to homeland security as well as the personal and financial security of citizens. It is not easy to gauge the amount of identity fraud at this moment in time. However, the minimum cost to the economy in some countries is in excess of $40bn per annum according to some official studies (see for example: Federal Trade Commission report released in 2004; UK Cabinet Office report released in 2002).

According to Gartner's recent study, about 15 million Americans were victims of fraud that stemmed from identity theft in the period from mid-2005 to mid-2006 (6). This represented an increase of more than 50 per cent from the reported 9.9 million in 2003 by the Federal Trade Commission. Current research and studies refer to the advances and spread of computer technology as the main factor behind this dramatic increase in identity theft (7). The literature shows that identity theft and fraud levels are increasing throughout the world (e.g., Canada, Australia, Britain, and Japan) with gigantic costs to victims and business (8). Some countries have introduced identity theft legislation that recognises such crimes and allows penalties and additional prison sentences (8).

However, countries around the world are realising that the legislation in itself cannot prevent or combat identity theft unless they adopt more effective and advanced solutions. One of the approaches pursued by many organisations both in government and private sectors is the employment of advanced technologies such as smart cards, biometrics, and PKI. It is widely argued that if properly implemented, these technologies can provide secure and accurate identity verification, enhance the security of the system and protect the integrity and confidentiality of information. The next few sections will look at these three technologies and explore them in further detail.

2. Biometrics

Biometrics* is defined as the science of using an individual's unique physical, behavioural and biological qualities for identification purposes e.g., fingerprint, hand print, facial recognition, iris, voice pattern, and so on. The first modern biometric device was introduced commercially over 20 years ago. Apart from being non-transferable among individuals, biometrics do not provide data about the person, but information.

The biometric industry has found a global market through smart card technology. Biometric identity cards are being adopted in many countries around the world. Analysts predict biometrics to boom in the next few years, referring to the recently released report from the International Biometric Group (IBG),** which indicated that the global market sales of biometric technologies would grow from less than $1bn in 2003 to more than $4.6bn in 2008, with fingerprint scanning becoming the most dominant technology, as illustrated in Figure 6.1 below. Governments and businesses are increasingly adopting biometric technologies in the belief that they will make identity theft and multiple identities impossible.

---

* The term 'biometrics' is derived from the Greek words bio (life) and metric (to measure).

** This is the biometrics industry's leading independent integration and consulting firm, providing a broad range of services to government and private sector clients.

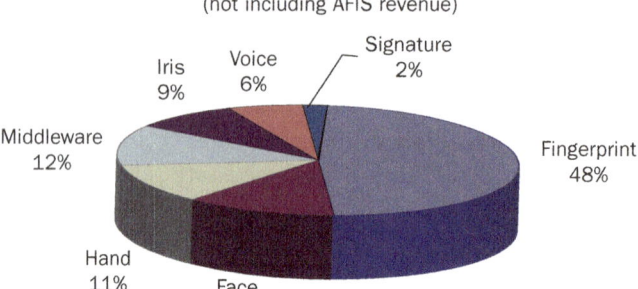

2004 Comparative market share by technology

(not including AFIS revenue)

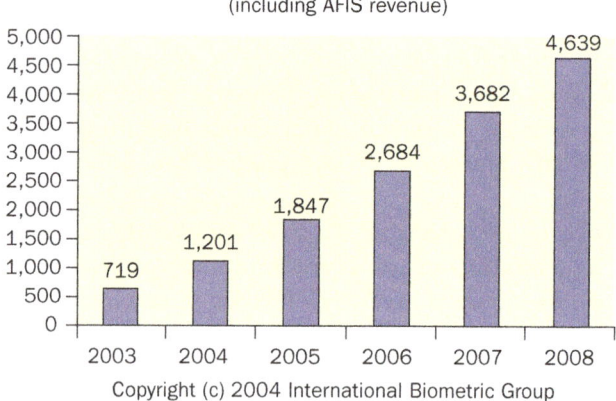

Total biometric revenue 2003 – 2008 ($m)

(including AFIS revenue)

Figure 6.1   Biometrics growth [9]

The National Physical Laboratory* conducted a performance evaluation test of several biometric technologies for a scenario of positive identification involving the following biometrics: face, fingerprint, hand geometry, iris, vein and voice recognition.

---

* The National Physical Laboratory (NPL) is the UK's national standards laboratory, an internationally respected and independent centre of excellence in research, development and knowledge transfer in measurement and materials science.

Iris recognition had the best accuracy, with 1.8 per cent false rejections and no false matches in over two million comparisons as illustrated in Figure 6.2.*

Of the other systems, fingerprinting performed best for low false acceptance rates (FAR),* while hand geometry achieved low (below one per cent) false rejection rates (FRR).** The study demonstrated that there is not yet one universal 'best' biometric system for both identification or authentication, but rather a combination of two or more biometrics may enhance the FAR and FRR factors (11).

In another evaluation, the UK National Physical Laboratory prepared a comprehensive feasibility study using biometrics as a means of

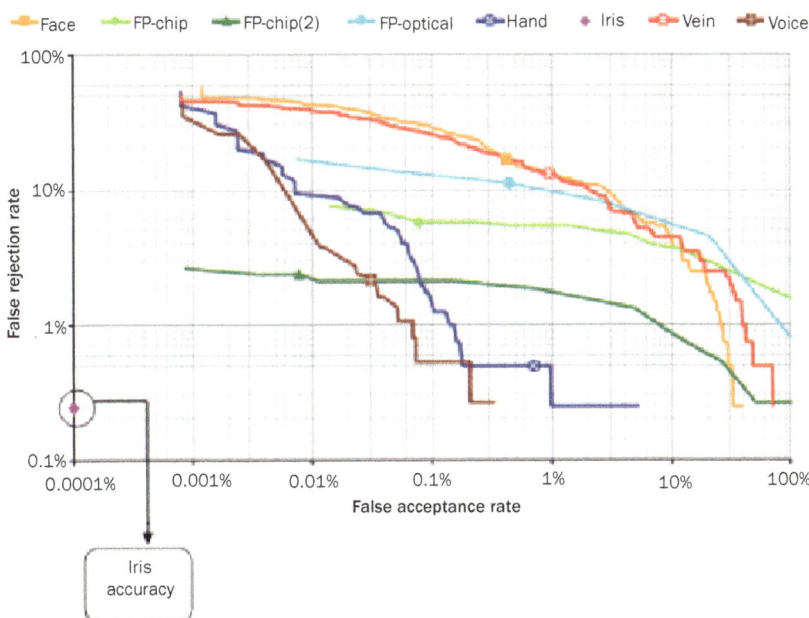

Figure 6.2 National physical lab results: FAR vs. FRR [12]

---

* False Acceptance Rate (also referred to as False Match Rate): incorrect identification or failure to reject an imposter (person trying to submit a biometric in either an intentional or inadvertent attempt to pass him/herself off as another person who is an enrollee).

** False Reject Rate (also referred to as False Non-Match Rate): failure to identify or verify a person

establishing unique identity to support the proposed entitlement scheme in development by the UK Passport Service and Driver and Vehicle Licensing Agency (12). The purpose of the study was to assess the feasibility of three main biometrics, namely fingerprint, iris, and face recognition technologies as a medium of identification in a national identity scheme, assessing the associated risks, and forwarding recommendations.

The feasibility study concluded once again that biometric methods do not offer 100 per cent certainty of authentication of individuals and that the success of any deployed system using biometric methods depends on many factors such as the degree of the 'uniqueness' of biometric measure, technical and social factors, user interface, and so on. However, and in principle, fingerprint and iris recognition were found to provide the identification performance required for unique identification over the entire UK adult population.

In the case of fingerprint recognition, the system required the enrolment of at least four fingers, whereas for iris recognition both iris were required to be registered. However, the practicalities of deploying either iris or fingerprint recognition in such a scheme were found to be far from straightforward in terms of the complexity of implementation, user training, and so on. Other studies show that it is the device and the algorithm used that actually determine the effectiveness of the biometric in use. A recent study by the National Institute of Standards and Technology (NIST)* revealed that fingerprint identification systems have 99 per cent accuracy with some enhanced devices and, perhaps more importantly, a slim 0.01 false positive rate i.e., only about one in 10,000 scans resulting in a misidentification (13). The study tested 34 fingerprint ID systems from 18 companies with about 50,000 sets of fingerprints from 25,000 people. The best systems reached 98.6 per cent accuracy for a single-print match, whereas two-finger matches were accurate 99.6 per cent of the time.

---

* NIST is a non-regulatory federal agency within the US Commerce Department's Technology Administration, with the mission to develop and promote measurement, standards, and technology to enhance productivity, facilitate trade, and improve the quality of life.

3. Smart cards

The 'smart card' is a plastic card with an IC (integrated circuit) chip capable of storing and processing data that may come with optional magnetic strips, bar codes, optical strips and holograms on a variety of card bodies. Developed in 1973 by the Frenchman Roland Marino, the smart card was not introduced commercially until 1981, when the French State telephone system adopted it as an integral part of its phone card network. This led to widespread use in France and then Germany, where patients have had health records stored on the cards. Table 6.1 provides a highlight on the developments of the smart card from the 1970s up to today.

Due to their capabilities, they are increasingly popular in many industries around the world, most particularly in telecommunications, but also banking, transportation (e.g., vehicle registration and driving licences) healthcare, insurance, and e-governance. With the increasing need for security, smart cards are being viewed as the ideal medium for implementing a secure identification and authentication infrastructure (14, 15). Smart card chips normally look like in the diagram below (Figure 6.3), with an integrated circuit built into it. This integrated circuit

Table 6.1　Smart card developments

1970s	Smart card technology invented by one of the Schlumberger companies (Axalto) to curb fraud
1980s	First commercial applications as a pre-paid memory card in the public telephony sector, followed by the banking industry which incorporated microprocessor capabilities
1990s	Telecommunication industry adopted smart cards as SIM cards
Mid 1990s	Advent of open platform cards e.g., Java cards (invented in 1996) which boosted multi-application cards, use of complex cryptography and becoming a medium to store, carry and transact with digital signatures. Introduction of contact-less technology and the invention of combi cards (contact + contactless) in 1996–97
Late 1990s	Applications based on contact-less technology and the invention of combi cards (contact + contactless) in 1996–97
2002	Invention of .NET technology in 2002 which led to the increase of smart card memory capacity to 512 K byte

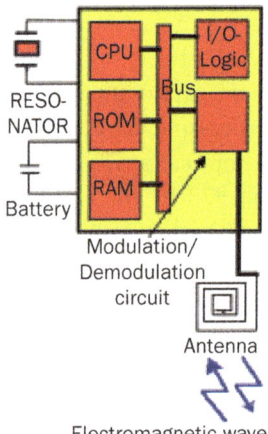

Electromagnetic wave

Contact Cards	The most widely used. They have to be moved past a reader i.e., require insertion into a smart card reader with a direct connection to a conductive micro-module on the surface of the card.
Contactless Cards	Require only close proximity (a few inches) to a reader.
Combi Cards	Could be used in both situations. Their main attraction is that one card could fulfill many purposes, such as credit card, bank card, membership card, ID-card and so on, all on in the same card.

Figure 6.3 **Types of smart card**

may consist only of EEPROM* in the case of a memory card, but may also contain ROM, RAM and a CPU.

As memory capacity, computing power, and data encryption capabilities of the microprocessor increase, many research studies indicate that smart cards are envisioned as replacing commonplace items such as cash,

---

* (Electrically Erasable Programmable ROM) A memory chip that holds its content without power. It can be erased, either within the computer or externally and usually requires more voltage for erasure than the common +5 volts used in logic circuits. It functions like nonvolatile RAM, but writing to EEPROM is slower than writing to RAM. EEPROMs are used in devices that must keep data up-to-date without power. For example, a price list could be maintained in EEPROM chips in a point of sale terminal that is turned off at night. When prices change, the EEPROMs can be updated from a central computer during the day. EEPROMs have a lifespan of between 10K and 100K write cycles. Source: *http://www.gurunet.com.*

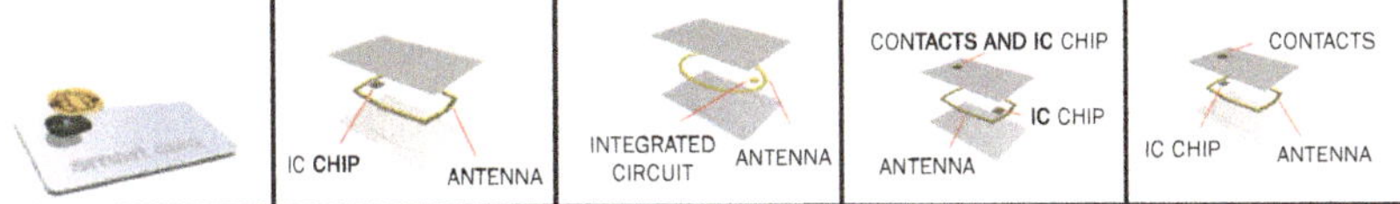

Contact Cards	Contactless Cards	Proximity Cards	Hybrid Cards	Combi Cards
IC CHIP	ANTENNA	INTEGRATED CIRCUIT ANTENNA	CONTACTS AND IC CHIP / IC CHIP / ANTENNA	CONTACTS / IC CHIP ANTENNA
Cards the size of a conventional credit or debit card with a single embedded integrated circuit chip that contains just memory or memory plus a microprocessor	Cards containing an enbeddes antenna instead of contact pads attached to the chip for reading and writing information contained in the chip's memory.	'Proximity cards' communicate through an antenna similar to contactless smart cards except that they are read-only	Cards containing two or more embedded chip technologies such as a proximity chip with its antenna and a contact smart chip with its contact pads.	Cards containing one smart chip thar can be accessed through either contact pads or an embedded antenna.
Popular Uses: Network security , vending, meal plans, loyalty, electronic cash, government IDs, campus IDs, e-commerce, health cards	**Popular Uses:** Student identification, electronic passport, vending, parking, tolls, Ids	**Popular Uses:** Security, identification, access control	**Popular Uses:** Accommodates legacy system infrastructure while adding applications that require different e-card technologies	**Popular Uses:** Mass transit and access control combined with other applications such as network security , vending , meals plans , loyalty

Figure 6.4   Emerging card technologies

airline and theatre tickets, credit and debit cards, toll tokens, medical records, and keys. Figure 6.4 provides further information about the emerging card technologies and their uses.

Smart cards are being widely adopted in many e-government and e-business initiatives as a vital element of a secure identification infrastructure and as a platform to hold both biometrics and PKI (16). The next section will explain the PKI and its role in providing a more highly trusted standard of authentication.

4. Public key infrastructure

PKI is a framework for creating a secure method for exchanging information based on public key cryptography. It is widely considered to be one of the prime components along with smart card and biometric technologies to enhance the overall security of systems. PKI is known to provide two main features:

(a) security, and (b) encryption, to fulfil four vital requirements and establish what is called a trust environment (see also Figure 6.5):

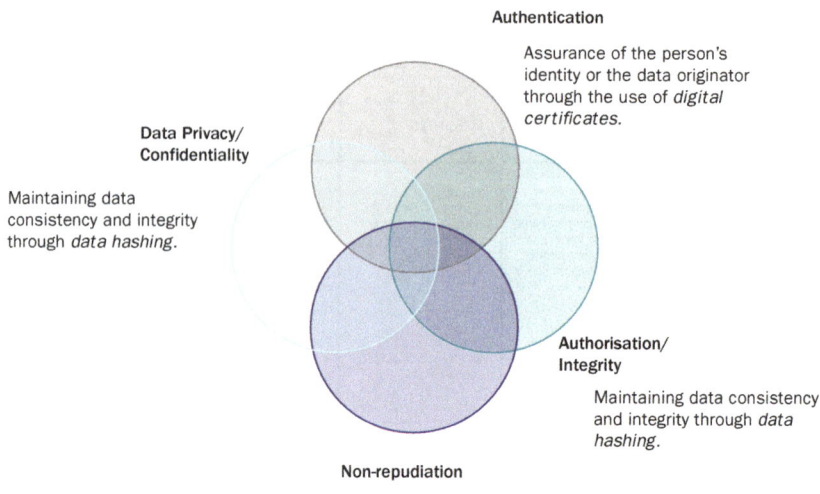

Authentication

Assurance of the person's identity or the data originator through the use of *digital certificates.*

Data Privacy/ Confidentiality

Maintaining data consistency and integrity through *data hashing.*

Authorisation/ Integrity

Maintaining data consistency and integrity through *data hashing.*

Non-repudiation

Communication originator can't deny it later i.e., guarantees the ownership of the electronic document through the use of the use of *digital signatures.*

Figure 6.5   **PKI trust framework**

(1) Authentication.
(2) Confidentiality.
(3) Integrity.
(4) Non-repudiation.

In a PKI environment, a pair of two different cryptographic keys is used for encryption and decryption purposes, referred to as public and private keys. The private key is kept secret by the user or in the system, and the public key is made public. The keys are mathematically related and could not be deducted from one another. Data encrypted with one key can be decrypted only with the other complementary key, and vice versa (see also Figure 6.6).

PKI encompasses a set of complex technologies as illustrated in Table 6.2 which shows the main PKI components. In a PKI environment, one would require a digital certificate, which usually contains the individual's public key, information about the certificate authority, and additional information about the certificate holder. The certificate is created and signed (digital signature) by a trusted third party, a certificate authority (CA). The individual's identity is bound to the public key, where the CA takes liability for the authenticity of that public key, to allow a secure communication environment.

public key   private key

plaintext   encryption   ciphertext   decryption   plaintext

Figure 6.6 PKI trust framework

Table 6.2 PKI architecture

Security policy	• Defines requirements and standards for the issue and management of keys, certificates and the obligations of all PKI entities, and is used to determine level of trust the certificate affords
Certification Authority (CA)	• Authenticates subscribers, issues and manages certificates, schedules expiry date for certificates and revokes them when the validity period expires.
Registration Authority (RA)	• Provides the interface between the user and CA. It verifies the identity of the user and passes the valid requests to the CA.
Certificate distribution system	• Usually through a directory service. A directory server may already exist within an organisation or may be supplied as part of the PKI solution.
PKI Enabled applications	• A cryptographic toolkit employed to enable PKI applications e.g., communications between web servers and browsers, email, electronic data interchange (EDI), virtual private networks (VPNs), and so on.

The registration authority (RA) is where the individual or the organisation requesting the certificate is checked to ensure that they are who they claim they are. Another fundamental component of PKI is the certificate distribution system which publishes the certificates in the form of an electronic directory, database, or through an email to allow users to find them. PKI-enabled applications usually refer to applications that have had a particular CA software supplier's toolkit added to them so that they are able to use the supplier's CA and certificates to implement PKI functions such as in emails and networks for encrypting messages. The certificate policy, also referred to as certificate management system, is where the certificate procedures are defined including the legal liabilities and responsibilities of the involved parties.

A digital signature is based on a range of encryption techniques; one of the essential services of PKI allows people and organisations to electronically certify such features as their identity, their ability to pay, or the authenticity of an electronic document. The digital signature, also referred to as an encrypted hashed text, is a digital fingerprint; a value which is calculated from the information in a message through the use of a cryptographic hash function. Any change to the message, even of a single bit, typically results in a dramatically different message.

Figure 6.7 shows an example of a system generating a hash value from the message and encrypting it with the originator's private key. The message, which could also be encrypted, is sent along with the digital signature to the recipient who will then decrypt the digital signature with the sender's public key to change it back into a message.

If the decryption is successful then it proves that the sender has signed the message, because only he/she has the private key. The recipient then calculates the hash value out of the received message, and compares it with the actual message. If the message is the same as the message created when the signature was decrypted, then the receiver can be assured that the signed message/data has not been changed or tampered with.

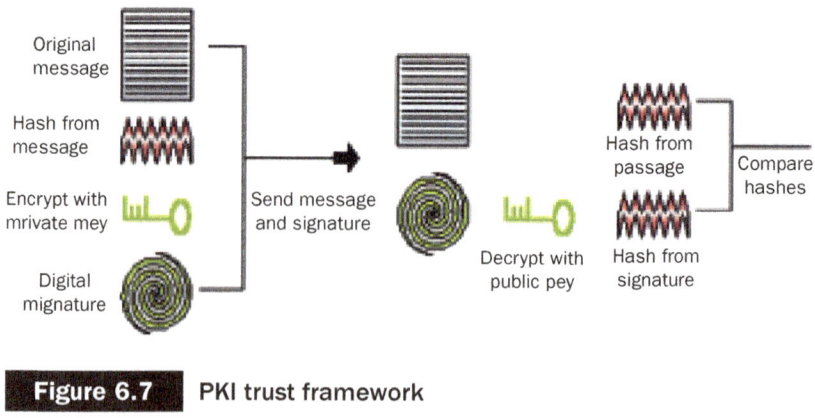

Original message

Hash from message

Encrypt with mrivate mey

Send message and signature

Digital mignature

Decrypt with public pey

Hash from passage

Hash from signature

Compare hashes

Figure 6.7   **PKI trust framework**

In their study to understand PKI infrastructure and how it may support electronic authentication and e-governments, (17) adopted an organisational framework to facilitate the understanding and classification of electronic services according to their security requirements (e.g. issuing birth certificates, submitting tax forms, conducting electronic payments, and so on).

The findings of the study demonstrated that the security services offered by the public key infrastructure can be employed for fulfilling most of the identified security requirements for an integrated e-authentication platform and a one-stop e-government portal as illustrated in Table 6.3. However, other requirements like availability, performance, un-coercibiliy, un-traceability, and anonymity could not be fulfilled, and additional security measures were found necessary.

In addition, several studies have proved that PKI is state of the art technology in the field of digital authentication and overall security infrastructure (17, 18, 19). Nonetheless, studies also show that PKI on its own will not provide maximum security for authentication unless it is incorporated with other security technologies such as smart cards, biometrics, virtual private networks, and so on (20, 21, 22).

Table 6.3 Use of PKI services for fulfilling e-government security requirements (17)

PKI services	Security requirements Availability	Performance	Management of privileges	Authentication	Logging	Integrity	Confidentiality	Non-repudiation	Anonymity	Public Trust	Untraceability	Secure storage
Registration			☑	☑					☑[1]			
Digital Signatures				☑		☑		☑				
Encryption							☑					☑
Time Stamping					☑			☑		☑		☑
Non-repudiation								☑				☑
Key Management				☑		☑	☑	☑				☑
Certificate Management				☑		☑	☑	☑				☑
Information Repository								☑		☑		☑
Directory services				☑		☑	☑	☑				☑
Camouflaging Communication							☑		☑[1]		☑[1]	
Authorisation			☑	☑								
Audit								☑		☑		☑
TTP to TTP interoperability				☑		☑				☑		

[1] Not in the context of e voting.

5. The application of the technology trio

As explained earlier, statistical data in the literature provides horrifying data about identity theft and how much it is costing both public and private organisations. In order to combat identity theft, organisations need the means by which they can accurately recognise peoples' identities in two main forms as illustrated in Figure 6.8:

1. Identification (1.N).
2. Verification – also referred to as authentication - (1:1).

The technology trio of PKI, smart cards and biometrics is being widely considered to address the need for precise identification and authentication of individuals. They offer a solid business model that not only addresses high-level security requirements and strong authentication but also protects individual privacy and preserves resources. The adoption of these three technologies will create two fundamental elements:

1.   A reliable mechanism to identify and authenticate individuals.
2.   A secure communication and transactional environment.

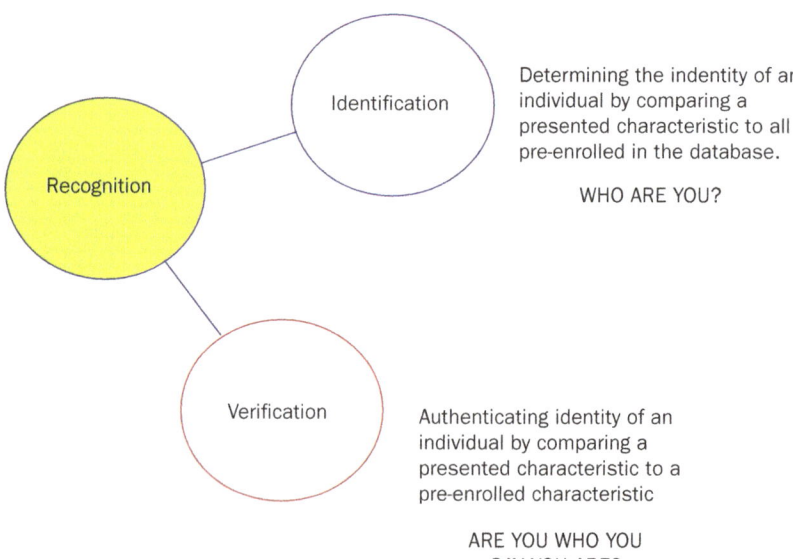

Figure 6.8   **Accurate identification/verification requirements**

Smart cards, for instance, can serve as the issuer's agent of trust and deliver unique capabilities to securely and accurately verify the identity of the cardholder, authenticate the ID credential, and serve the credential to the ID system (23). PKI, on the other hand, has emerged as the most reliable framework for ensuring security and trust (24, 25).

Apart from the main benefit of PKI in enabling secure electronic transactions, it can also be used to encrypt the data stored in the chip (e.g., personal information, digital photo, biometrics, and so on), in addition to the data stored in the database, to limit access to only authorised persons and entities.

Biometrics allows the padlocking of the person to the card. In doing so, the card cannot easily be transferred to another individual. In particular, given the current focus on the use of biometrics in ID card systems, it sets out architecture for strongly-authenticated identity cards that deliver (perhaps counter-intuitively) both enhanced security and enhanced privacy.

Through the incorporation of these three technologies in an identity management system, individuals are not locked into one form of authentication, but rather three different forms of authentication (see Figure 6.9):

1. *Knowledge factor:* a password to ascertain what one knows.
2. *Possession factor:* a token (smartcard) to ascertain what one has.
3. *Biometric factor:* biometric recognition (for example fingerprint or thumbprint) to ascertain who one biologically is.

As such, if one factor has been compromised, fraudsters need to pass through another two levels of authentication. By requiring three forms

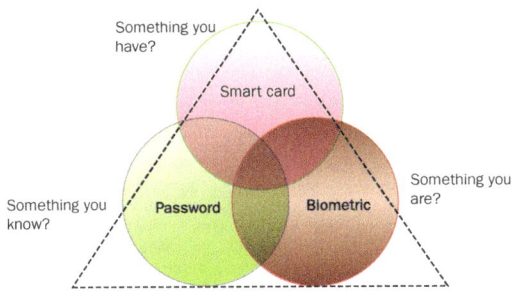

Figure 6.9   Three factor authentication

of identification to access credentials, organisations will be able to bind card holders' (digital) identities on the card to their physical identities.

From an e-government perspective, the key to G2C e-government is authentication i.e., the ability to positively identifying and proving the authenticity of those with whom the government conduct business with. Without authentication, other security measures put in place for many G2C transactions can be ineffective.

The argument here is that for a G2C e-government to progress, governments need a strong online trusted authentication infrastructure, without which, their efforts are likely to be at a stand still. In other words, governments need varying levels of authentication strength based on the value or sensitivity of their online information or services, balanced against other considerations like usability, deployment, and budget.

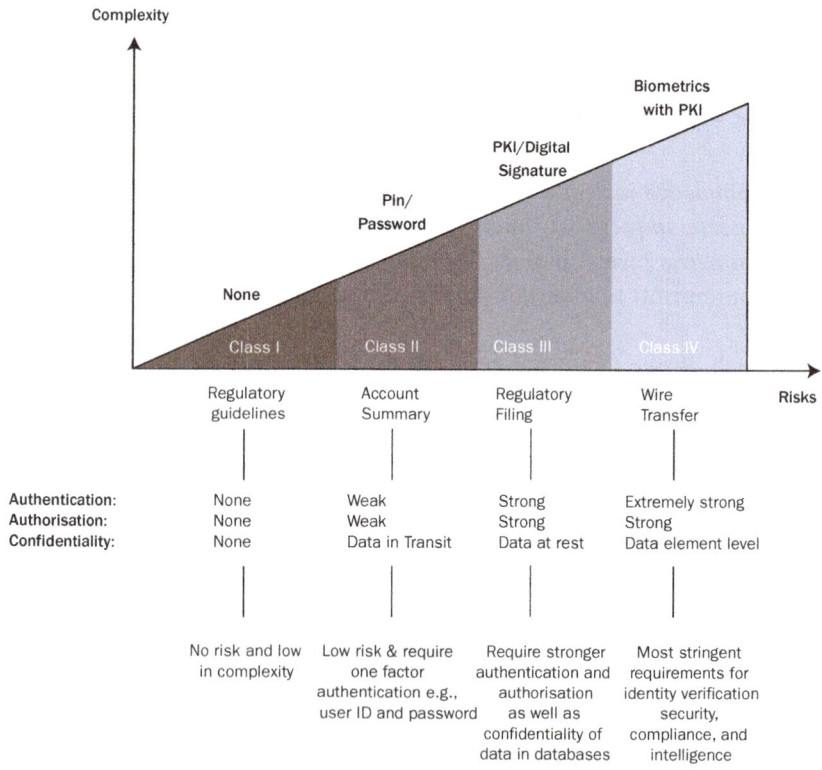

Figure 6.10 An example of types of authentication for G2C e-government services

It is important to heed that the essence of G2C e-government is that transactions occur between people that are represented by machines. The anonymity of these transactions makes it more difficult to identify the parties involved and to ensure a trusted business relationship. Since all successful business relationships are based on trust, establishing online trust should be one of the primary goals of any e-government initiative (23).

The focus must be on building a trusting environment that provides a high level of data privacy, data integrity, and user authorisation. Nonetheless, and as mentioned earlier, the real cornerstone of G2C e-business trust is authentication: that is, knowing with whom the government is doing business.

PKI, smart cards, and biometrics are the technologies that are believed to be the key components of the trust model to address both electronic transactions' security and online identity authentication. Using the power of the presented technologies in this article, government organisations and businesses alike can use varying levels of authentication depending on the level of security required for a particular transaction as depicted in Figure 6.10.

6. Conclusion

Organisations will be better able to protect their systems and assets with the application of biometrics, smart cards, and PKI, which provides them with a better verification of both physical and virtual identities. On the plus side, the principal advantage to be gained is more reliable authentication of identities. Without a doubt, strong user authentication must be viewed as the foundation for any e-government and e-commerce initiatives (17). In fact, apart from improving traditional approaches to identification and authentication, these technologies are seen as the key to e-government, and a secure digital infrastructure.

This utilisation of the three named technologies in this paper should have a profound positive impact, not only in terms of the reduction of identity theft and fraudulent activities, but having such an infrastructure should enable the improvement of current government services and pave the way for more investment in electronic services. In short, the promise of the technology trio is colossal. Hopefully, future applications and developments, implemented in well managed products, will prove this right.

References

1. Briefel, A. (2006) 'The Deceivers: Art Forgery and Identity in the Nineteenth Century'.Cornell University Press.
2. Shute, J. (2006) 'User I.D.: A Novel of Identity Theft. Mariner Books'.
3. Adams, J. (2003) 'E-Fraud Fight Prompts Credit Agencies' Cooperation'. *Bank Technology News* 16 (6): 14.
4. Marcus, R. and Hastings, G. (2006) 'Identity Theft'. Disinformation Company Inc.
5. Middlemiss, J. (2004) 'Gone Phishing'. *Wall Street & Technology*, August: 38-39.
6. McCarthy, C. (2007) 'Study: Identity theft keeps climbing'. *CNET News. com. http://news.com.com/21001029_3616476-5.html*
7. Zalud, B. (2003) 'Real or fake'? *Security* 40 (3): 12-18.
8. Anonymous (2003) 'ID theft tops fraud list again'. *ABA Bank Compliance* 2: 5–6.
9. International Biometric Group, *www.biometricgroup.com*
10. Mansfield, T., Kelly, G., Chandler, D. and Kane, J. (2001) 'Biometric Product Testing – Final Report'. National Physical Laboratory: UK, *http://www.cesg.gov.uk/site/ast/biometrics/m- edia/BiometricTestReportpt1.pdf*
11. Mansfield, T. (2001) 'Biometric Authentication in the real world'. National Physical Laboratory, UK. *http://www.npl.co.uk/scientific_software/publications/biometrics/psrevho.pdf*
12. Mansfield, T. and Rejman-Greene, M. (2003) 'Feasibility Study on the Use of Biometrics in an Entitlement Scheme for UKPS, DVLA and the Home Office'. National Physical Laboratory, UK: *http://uk.sitestat.com/homeoffice/homeoffice/s?docs2.feasibility_study031111_v2&ns_type=pdf*
13. McCeary, L. (2004) 'The Fact of Fingerprints, The Resource for Security Executives'. *http://www.keepmed-ia.com/pubs/CSO*
14. George, T.C. (2003) 'The inside of a smart story: smart cards are increasingly becoming relevant in our everyday life'. *Businessline*, Chennai, October: 1.
15. MacGowan, J. (2003) 'Smart Cards: Enabling e-Government', Bloor Research, *http://www.itanalysis.com/art-icle.php?articleid=11151.*
16. Gardner, S. 2004 'Europe Get Smart'. *www.euro-correspondent.comwww. eurocorrespondent.com/ed60_110-704.htm*
17. Lambrinoudakis, C., Gritzalis, S., Dridi, F., and Pernul, G. (2003) 'Security requirements for e-government services: a methodological approach for developing a common PKI-based security policy'. *Computer Communications* 26: 1873–1883.
18. Conry-Murray, A. (2002) 'PKI: Coming to an enterprise near you?' *Network Magazine* 17 (8): 34–37.
19. Critchlow, D. and Zhang, N. (2004) 'Security enhanced accountable anonymous PKI certificates for mobile e-commerce'. *Computer Networks* 45 (4).
20. Ellison, C. and Schneier, B. (2000) 'Ten Risks of PKI: What You're not Being Told about Public Key Infrastructure'. *Computer Security Journal* 16 (1): 1–8.

21. Doan, (2003) 'Biometrics and PKI based Digital Signatures'. White Paper, *Daon, www.daon.com.*
22. Kolodzinski, O.(2002)'PKI: Commentary and observations'. *The CPA Journal* 72 (11): 10.
23. SCA (2004) 'Secure Identification Systems: Building a Chain of Trust, Smart Card Alliance'. Available at: *http://www.smartcardalliance.org.*
24. Hutchison, R. (2000) 'E-Government: Walk before you run'. *Canadian Business* 73 (16): 36.
25. Russell, S., Dawson, Ed., Okamoto, E. and Lopez, J. (2003) 'Virtual certificates and synthetic certificates: new paradigms for improving public key validation'. *Computer Communications* 26: 1826–1838.
26. *http://www.idedge.com*
27. Al-Khouri, A.M. and Bal. J. (2007) 'Electronic Government in the GCC countries'. *International Journal of Social Sciences* 1 (2): 83–98.

Iris recognition and the challenge of homeland and border security in the UAE

Ahmad N. Al-Raisi and Ali M. Al-Khouri

Abstract: This article discusses the implementation of iris recognition in improving the security of border control systems in the United Arab Emirates. The article explains the significance of the implemented solution and the advantages the government has gained to-date. The UAE deployment of iris recognition technology is currently the largest in the world, both in terms of number of iris records enrolled (more than 840,751) and number of iris comparisons performed daily 6,225,761,155 (6.2 billion) in 'all-against-all' search mode.

Key words: *Border control, homeland security, biometrics, iris recognition.*

1. Introduction

Today security has become a top priority subject on many countries' agendas, as governments find themselves faced with continuous radical strategic challenges related to identity management and verification. Despite the fact that they will not be a panacea in every case, biometric technologies are at the forefront of these agenda discussions since they provide a highly accurate identity confirmation which makes it seen as a very effective answer to many security and identity management impediment issues. Recent advances in technology, coupled with a significant price drop, and fuelled by the legislative requirements for positive identification and verification, the biometric industry is growing enormously with an ever-increasing market share as a viable alternative to upgrade security levels in local, regional and national security checkpoints.

The term *biometrics* refers to a wide range of technologies available in the market to identify and verify a person's identity by means of measuring and analysing various human physiological and behavioural characteristics. In order to make a decision of which biometric product or combination of products would satisfy the stated requirements, different factors need to be assessed. Factors for consideration would typically include accuracy of a specific technology, user acceptance, and the costs of implementation and operation.

Table 1 summarises some of the important biometric features that need to be taken into account when comparing different biometric technologies. The iris is seen as a highly reliable biometric technology

Table 7.1   Important features of biometric technologies

Technology Characteristic	Fingerprint	Iris	Facial	Hand
How it works	Captures and compares fingertip patterns	Captures and compares iris patterns	Captures and compares facial patterns	Measures and compares dimensions of hand and fingers
Cost of device	Low	High	Moderate	Moderate
Enrolment time	About three minutes, 30 seconds	two minutes, 15 seconds	About three minutes	About one minute
Transaction time[a]	nine to 19 seconds	12 seconds	ten seconds	six to ten seconds
False nonmatch rate[b]	.2%–36%	1.9%-6%	3.3%–70%	0%–5%
False match rate (FMR)[c]	0%–8%	Less than 1%	0.3%–5%	0%–2.1%
User acceptance issues	Associated with law enforcement, hygiene concerns	User resistance, usage difficulty	Potential for privacy misuse	Hygiene concerns
Factors affecting performance[d]	Dirty, dry, or worn fingertips	Poor eyesight, glare, or reflections	Lighting, orientation of face, and sunglasses	Hand injuries, arthritis, swelling
Demonstrated vulnerability[e]	Artificial fingers, reactivated latent prints	High-resolution picture of iris	Notebook computer with digital photographs	None
Variability with ages[f]	Stable	Stable	Affected by aging	Stable
Commercial availability since	1970s	1997	1990s	1970s

Source: Dillingham (2002)

because of its stability, and the high degree of variation in irises between individuals. The discussion here in this article is limited to the iris as the next sections will explore it in more detail.

This paper is structured as follows. First, some introductory background information is provided about the iris and its characteristics. Then some accuracy and performance evaluation tests carried out to measure its accuracy are put forward to highlight the reported findings. The following sections mainly deal with the iris implementation in the UAE from its pilot to its mass roll-out phase. Some high level information is also presented about the system's architecture and the recent statistics from the UAE iris system. The paper prior to the conclusion presents some lessons learned and a number of provisioned applications of iris technology in the future.

2. Background to iris recognition

Identifying a person from his/her iris record was a concept first thought of by Drs. Safir and Flom, American ophthalmologists (Flom and Safir, 1987). The algorithm underlying iris recognition to read and map the data in a person's iris was developed by Dr. John Daugman, a Harvard PhD graduate and a noted computer scientist at Cambridge University in England.

The US Patent 5,291,560 issued in the name of Daugman (Daugman, 1994) has been assigned to Iridian Technologies, Inc., one of the world's principal vendors of iris-based systems to hold the exclusive worldwide patents on iris recognition (Heath, 2001). The iris pattern variability between different people is enormous. No two irises are alike. Unlike DNA or even fingerprints, iris recognition works by performing exhaustive searches to identify individuals in real time.

The iris (the coloured ring surrounding the pupil) has in excess of 266 mathematically unique characteristics. The retina on the other hand, is the

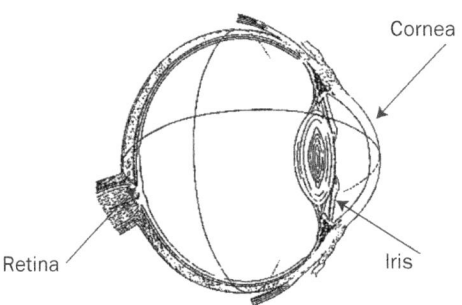

Figure 7.1 What is the iris?

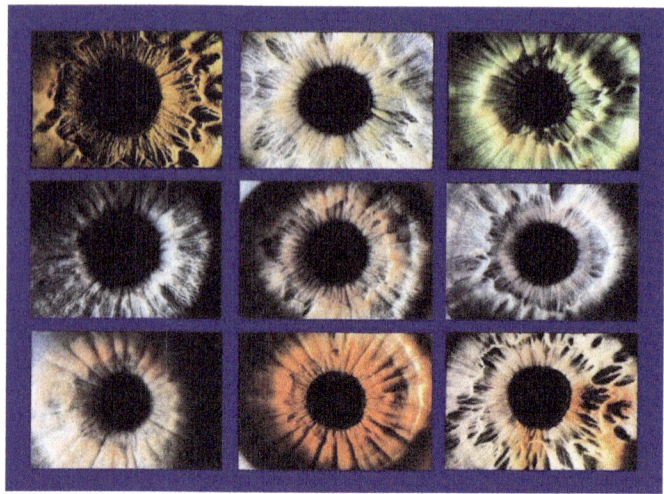

Figure 7.2 A collection of irises

hemispherical organ behind the cornea, lens, iris, pupil, and is not readily visible (see Figure 7.1). With no genetic influence on its development, the iris is permanently formed by the eighth month of gestation, a process known as 'chaotic morphogenesis' (Daugman, 1993).

In contrast to other biometrics such as fingerprints, the iris is seen as a highly reliable biometric technology because of its stability, and the high degree of variation in irises between individuals. Figure 7.2 demonstrates the variations found in irises.

The likelihood of iris damage and/or abrasion is minimal since it is protected by the body's own mechanisms i.e., it is behind the eyelid, cornea and aqueous humour. Extensive research has determined that the human iris is not subject to the effects of aging and it remains unchanged in structure and appearance from the time it is developed until a few minutes after death.

3. Accuracy and performance measurement

The accuracy of a biometric system is commonly measured by two factors; false acceptance rate (FAR) and false rejection rate (FRR). Also referred to as a *type II* error, FAR is considered the most serious biometric security error, as it represents the system incorrectly accepting an access attempt by an unauthorised user. On the other hand, FRR, also referred to as a *type I* error, is the measure of the likelihood that the biometric security system will incorrectly reject an access attempt by an authorised user.

A false rejection does not necessarily indicate a flaw in the biometric system. In a fingerprint-based system, for instance, an incorrectly aligned finger on the scanner or dirt on the scanner can result in the scanner misreading the fingerprint, causing a false rejection of the authorised user.

In general, there is typically a direct correlation between FAR and FRR. The lower the FRR percentage the higher the FAR percentage and vice-versa. Finding a medium that keeps both FAR and FRR to a minimum can be difficult. The degree of difficulty depends on the biometric method chosen and the vendor implementation.

For the reason that FAR and FRR are interdependent, it is important to determine the threshold values for these two factors (Liu and Silverman, 2001). Figure 7.3 plots the two factors against each other, where each point on the plot represents a hypothetical system's performance at various sensitivity settings. With such a plot, we can compare these rates to determine the crossover error rate, also known as the equal error rate (EER) (Liu and Silverman, 2001).

This value indicates the error rate at which proportion of FAR equals the proportion of FRR. The lower the equal error rate value, the higher the accuracy of the biometric system. Another factor that needs to be considered with regards to accuracy measurement is the score known as the *hamming distance* that is discussed next.

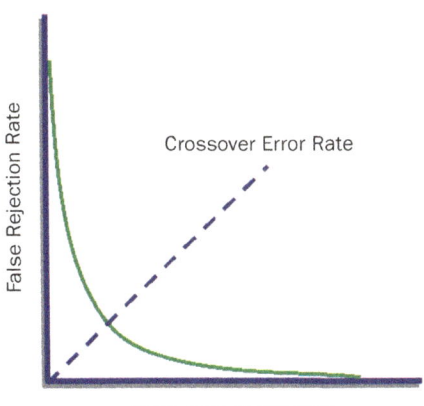

Figure 7.3 The crossover error rate attempts to combine two measures of biometric accuracy

Source: Liu & Silverman (2001)

3.1 Hamming distance

In order to measure the difference or 'variation' between two given IrisCodes™, the hamming distance (HD) is normally calculated. The way it works is that when compared against each other (bit by bit), if the two bits are identical, the system assigns a value of *zero* to that pair comparison, and *one* if they are different as illustrated in Figure 7.4. If the distance between the two compared iris codes is below a certain threshold, they are called a match. In other words, if two patterns are derived from the same iris, the hamming distance between them, in theory, will be close to 0.0 due to a high correlation.

The smallest hamming distance corresponds to the best match between two templates e.g., a hamming distance of 0.10 means that two IrisCodes™ are different by ten per cent. Furthermore, as is the case with any biometric, as one may force the threshold lower, the likelihood of a false rejection increases. In a comprehensive 200 billion cross comparisons carried out on the UAE Database by Prof. John Daugman of Cambridge University (discussed in section 10), not a single false match was found lower than the hamming distance of 0.262.

3.2 Performance tests

Several performance and evaluation tests over the last nine years have identified iris recognition technology as the most accurate biometric. Table 7.2 depicts a summary of the different independent performance

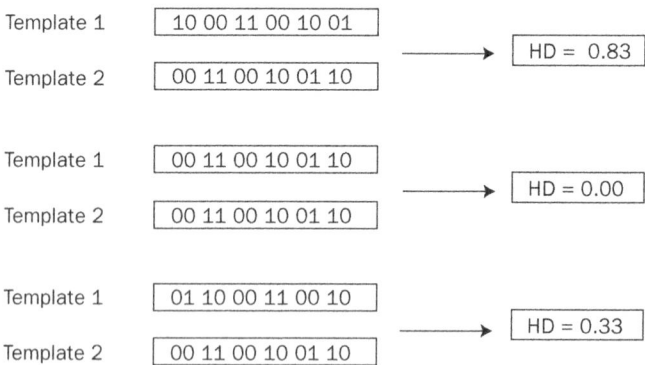

Figure 7.4 Iris matching and HD calculation

Source: (Daugman, 2005)

Table 7.2 Iris performance tests

Testing Body	Year	Comparisons	False Match
Sandia Labs, USA	1996	19,701	None
British Telecom Labs, UK	1997	222,743	None
Sensar Corp., USA	1999	499,500	None
Joh. Enschede, NL	2000	19,900	None
Prof. John Daugman, UK	2000	2,300,000	None
Eye Ticket, UK	2001	300,00	None
National Physical Labs, UK	2001	2,735,529	None
Prof. John Daugman, UK	2002	9,200,000	None
Iridian Technologies, USA	2003	984,000,000	None

tests performed since 1996 to measure the accuracy of the iris. The largest sample performed was by Prof. John Daugman in 2002, with around nine million comparisons and showed a surprising zero false match rate.

Another evaluation was reported by the National Physical Laboratory* in April 2001. The performance test included an evaluation of several biometric technologies for a scenario of positive identification involving the following biometrics: face, fingerprint, hand geometry, iris, vein and voice recognition as illustrated in Figure 7.5.

Iris recognition had the best accuracy, with 0.1 percent false rejections, no false matches in over two million comparisons and a 0.0% failure-to-acquire rate (Mansfield et al., 2001).

Of the other systems, the fingerprint performed best for low false acceptance rates (FAR), while hand geometry achieved low (below 1 per cent) false rejection rates (FRR). The study illustrated that there is no one universal 'best' biometric system yet, rather a combination of two or more biometrics may enhance the FAR and FRR factors (Mansfield, 2001).

---

* The National Physical Laboratory (NPL) is the UK's national standards laboratory, an internationally respected and independent centre of excellence in research, development and knowledge transfer in measurement and materials science.

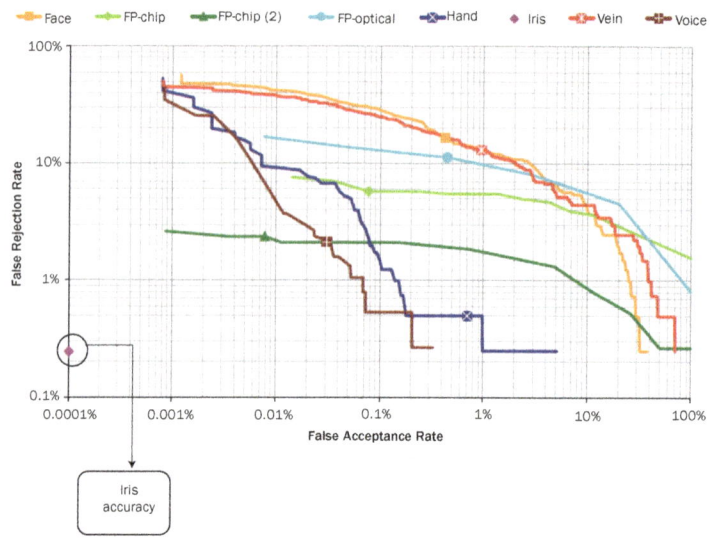

Figure 7.5 National physical lab results: FAR vs. FRR

Source: Mansfield & Rejman-Greene (2003)

4. The challenge at UAE borders

The border control system in the United Arab Emirates is comprehensively managed in accordance with strict pre-qualification and visa issuing processes. Border control forms a key aspect of controlling the various ports of entry and exit throughout the UAE.

The physical processes, although not fully integrated, are functioning adequately at the various ports of entry. One of the greatest challenges the country is faced with is the repeated attempts of former expellees to re-enter the country (foreign nationals expelled for various violations). Various control measures have been implemented to detect such cases. However, these measures appeared to be inadequate to control and detect the return of deportees back into the country. The analysis of the status quo revealed that despite the huge investment in information technology systems at the Ministry of Interior, there was a clear gap in the accurate identification of a deported person who is back in the country either using fraudulent or genuine travel documents.

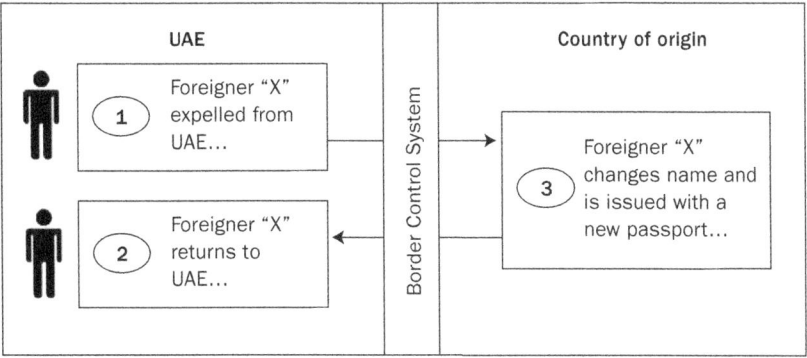

Figure 7.6   The challenge of returning expellees

Officials at the border used to rely on computer systems to check the validity of the presented documents and run a parallel data check against the blacklist database to check for a match. This showed a complete reliance on biographical data. The problem was that deported people were coming back into the country by changing their personal information such as name and date of birth and obtaining new passports that reflected these new details, which made the task of identifying him or her impossible.

Figure 7.6 shows a clear illustration of situations where a person is expelled from the country back to his country of origin. He changes his details and issues a new passport and returns back into the country, where the single point of failure is represented in the computer systems at borders leading to the existing control systems completely fail to detect such cases.

5. The technical solution

With extensive research and looking at the different lab results mentioned in section 3.2 above, the Information Technology Department at Abu Dhabi Police GHQ, which that was tasked to prepare a technical report on this matter, found out that biometrics would be a very effectual method to prevent illegal immigrants and former expellees from entering the country.

The desired biometric system was specified to be capable of scanning all incoming arrivals and provide positive or negative match feedback. The department then prepared a list of requirements that were later translated into specifications for the desired biometrics solution. Following are some criteria cited in the specification document and was used for evaluating the different biometric options:

- Can identify a single person from a large population of people.
- Does not change over time.
- Fast to acquire and easy to use.
- Can respond in real-time needed for mass transit locations (e.g., airports).
- Safe and non-invasive (disease control).
- Can be used in the millions and maintain top performance.
- Is affordable.

After extensive research and analysis of the different biometric products available on the market, iris recognition was found to satisfy most of the set requirements. Despite the newness of the technology at the time, the government decided to pilot iris recognition and have a pioneering global position in the implementation of such innovative technologies.

6. The pilot approach

The pilot implementation took around a year and a half in total. The first stage of the pilot operation involved requirements analysis, infrastructure setup, and system installation as well as the enrolment of expellees at three police stations with local databases. The build up of the iris database i.e., the acquisition process continued for around three months.

The three databases were then merged into a central database hosted at the Abu Dhabi police GHQ data centre as depicted in Figure 7.7. With only three iris cameras installed at Abu Dhabi international airport, the system was put to a real test.

Surprisingly, the system was successfully able to catch more than 50 people in less than four months of operation with a small number of registered expellees, indicating a serious threat to the existing border control system at the time. The result of the pilot was enough to get a

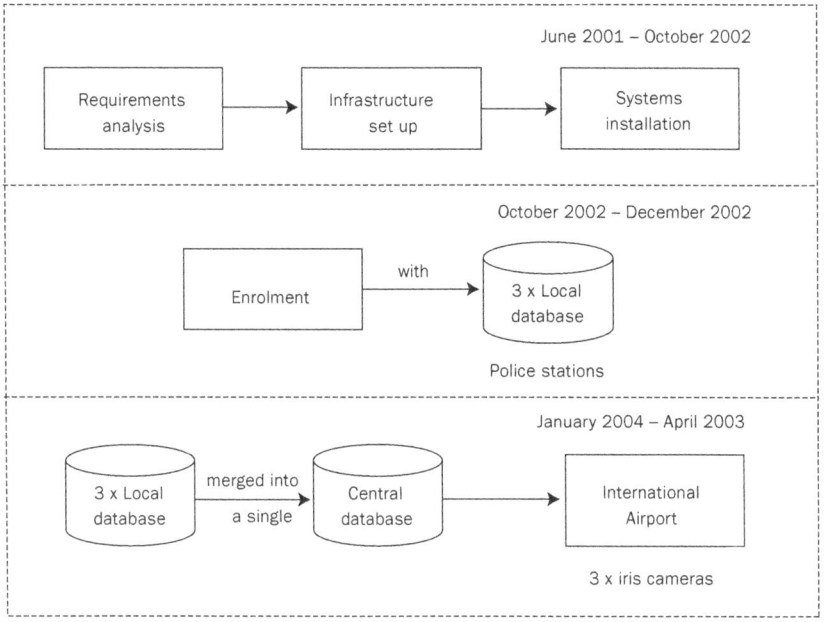

Result = 50 people caught in less than 3 months

Figure 7.7 Iris pilot in the UAE

buy-in from higher management and trigger a large-scale implementation of the system across the country.

7. Mass roll-out

The mass rollout of the system started in January 2003 and in less than five months, a total of 63 iris cameras were installed at 36 deportation (acquisition) centres and border control (recognition) points throughout the seven Emirates. The enrolment process involved the registration of inmates' and expellees' irises from geographically distributed prisons and deportation centres throughout the UAE into a central iris database.

Today, the UAE is considered to be the largest national deployment site of iris recognition in the world. More than 100 iris cameras are installed in all 17 air, land, and sea ports including the deportation centres with a total number of 20 enrolment centres and 27 border centres.

Via the secure national network infrastructure, each of the daily estimated 7,000 travellers* entering the country is compared against each of expellees, whose IrisCodes™ were registered in a central database upon expulsion. The real-time, one to all, iris check of all arriving passengers at any UAE border point reveals if the person has been expelled from the country.

It was again surprising sometimes how organised some criminal activities are. For instance, some cases showed that people, after being deported, come back to the country in less than 24 hours with genuine identity and travel documents with modified personal information such as name and date of birth. The latest statistics from the system show some absolutely amazing figures, indicating iris recognition effectiveness and the continuous attempts of those expelled to re-enter the country and challenge the system. The most recent statistics are presented in a later section in this paper.

8. The system architecture: how does it work?

The enrolment process which usually takes place in prisons or deportation centres distributed around the country, takes less than two minutes. The process involves the scanning of the person's irises, and storing them in the local database as shown in Figure 7.8.

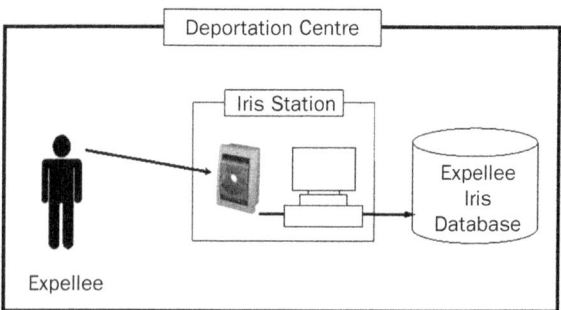

Figure 7.8   Expellee enrolment process

---

* Travellers required to go through an iris check are those who enter the country for the first time with a work permit entry visa, and those with visitor visas from certain non-visa waiver countries.

IrisCodes™ collected at enrolment centres around the country are deposited into a central iris repository database, which performs database management such linking with geographical and time base data, as well as update and maintenance. The local databases are synchronised with the central database with a user-set refresh call every two minutes (see also Figure 7.9).

8.1 System platform and components

The UAE system is constructed from a variety of Commercially Off The Shelf (COTS) components integrated with special Iris Technology cameras, interfaces and software. The databases of irises are all memory resident, offering unparalleled search speeds reaching more than 650,000 iris comparisons per second. The operating system upon which the system resides is the Microsoft Windows 2000 platform. All communications between the central repository and the geographically detached locations are encrypted TCP/IP communications (see also section 8.6).

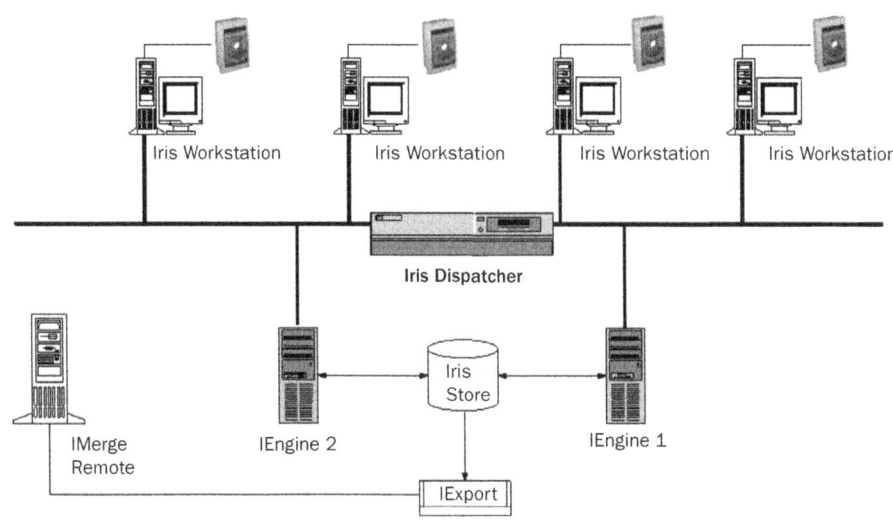

Figure 7.9   Typical deportation structure

8.2 System performance and scalability

A full enrolment process of a person does not take more than 30-45 seconds by a trained operator. This includes the enrolment of both irises and the typing of the associated biographical information. The overall search turn-around time does not exceed a few seconds.

It can perform over 650,000 iris searches per second. The system architecture is designed to sustain scalability without any loss of performance. The central iris repository can be served by an unlimited number of search engines each capable of searching with speeds exceeding 650,000 irises per second.

8.3 System threshold

All finds are reported at all border points as they are found. To assist the officer in interpreting the quality of the match, the following scale is used:

1. If the score is within 0.03 of the reported HD, the score is a *MATCH*.

2. If the score is within 0.03 to 0.04 of the reported HD, the score is a *HIGH MATCH*.

3. If the score is greater than 0.04 of the reported HD, the score is a *VERY HIGH MATCH*.

8.4 Synchronisation of the central database

As the enrolment process takes place in various enrolment centres, the central database makes routing polls to each enrolment centre and reads the latest set of information that may have been acquired by that centre since the last time the poll was carried out. This user-set, regular and fully automatic procedure ensures that the central database is always maintained in an up-to-date state. The synchronisation process is designed to be performed without any interruption of service or slowing down the performance of any of the enrolment or acquisition centres.

8.5 Application security

The system maintains a comprehensive audit trail that provides verification of all system-related activities e.g., records creation and modification, session logs, found records by each workstation, number of searches performed, and so on. Iris workstations are controlled by the central iris repository, which can disconnect remote workstations and redefine their search scope as well. Standard security and performance reports can be produced from the main site to monitor remote workstations at any time.

8.6 Data security

All IrisCodes™ in the system are encrypted using a TripleDES encryption system to provide maximum protection with a 192-bit key. All transmissions to and from the central database are also encrypted.

8.7 Fault tolerance

The system is designed to handle an unlimited number of enrolment centres and remote iris workstations working simultaneously. Each enrolment centre can lose the connection with the central iris repository without loss of function as each centre locally stores the IrisCode™ it enrols and, if a communication link is open, the balance of the recently enrolled irises will be transferred without user intervention.

Therefore, the enrolment centre can continue to enrol people independently of the availability of a communications link. For obvious reasons, the recognition sites need online connectivity to the central iris repository to perform the search operations. Although not yet available, the authorities are working to set up a disaster recovery site as a secondary backup central iris repository that will be synchronised with the main site. The enrolment sites will be configured to automatically switch over to the secondary site in case of failure of the primary one.

8.8 Back-up procedures

Back-ups are performed automatically at enrolment centres as well as the main site on backup tapes and hard disks. The total time for each back-up session does not exceed a few minutes. The system during this process will become temporarily unavailable, and the end-users are informed automatically through the status displayed on the iris workstations.

No user intervention is required in any location for the back-up to be carried out. The back-up procedure can be configured to occur multiple times depending on the requirement. The system supports a frequency setting of 0 (no back-up) up to 24 (once every hour).

9. Current UAE iris system statistics

As illustrated in Table 7.3, the UAE owns to date the largest iris database in the world with 840,751 iris records representing 152 different nationalities. The time required for an exhaustive search through the database is about three seconds.

So far around 6.5 million exhaustive searches against that database have been performed. The iris system in the UAE has performed more than 2.5 trillion comparisons* to date with a zero false match rate** under the 0.262 Hamming Distance.

On an average day, some 7,000 arriving passengers (peak of 12,142) are compared against the entire watch list of 840,751 in the database;

Table 7.3   UAE iris system statistics

Item	Value
Database size (IrisCode™):	840,751
Database size in bytes	2.8 Gbytes
New enrolments per day (rate of database growth per day)	700
Searches carried out between 2001 and 2005:	6,471,722
Average searches per day:	7,405
Daily cross comparisons (billion):	6.23 (6.23×10^9)
Expected next 12 months (trillion):	2.20 (2.2×10^{12})
Total comparisons to date (trillion):	2.5 (2.5×10^{12})
Persons caught:	56,484
Persons caught per day:	90-100
Search turn-around (including image acquisition):	3-4 Sec.
Date generated: 26 11 2005	

---

* Each system search involves one eye being searched and found or not found in the database. A cross comparison involves comparing one eye to the whole database of 840,751. So one search will represent 840,751 cross comparisons. The formula used to calculate the cross comparisons is to multiply the number of searches by the size of the database. So if we do 7,405 searches per day against a database of 840,751, the total daily cross comparisons performed becomes: 6,225,761,155 or 6.23 billion per day.

** If a match is found during the scanning of irises of incoming travellers at airports, the person is directed to another station connected to the immigration and black list system, capable of pulling the matching record which will typically show the person's particulars e.g., photo, name, expulsion data, crime, and so on. The authorities take action as appropriate.

this is about 6.5 billion comparisons per day with a sustained real-time response reported by all sites on a 24/7 basis. A total of 56,484 persons have so far been found on the watch list and seeking re-entry.

The number of searches is expected to rise considerably in the next few months as the government is currently working to include more traveller categories to submit for an iris recognition at all UAE border entry points. The amazing results of the system lie in the fact that 56,484 people were caught at borders attempting to re-enter the country after being deported using both forged and genuine travel documents. To the stakeholders, this means a great return on investment.

10. The UAE study

The UAE study was based on 632,500 IrisCodes™ acquired from the UAE system, representing 152 nationalities, where they were compared against each other generating over 200 billion comparisons in total as shown in Figure 7.10. The task that was performed by Prof. John

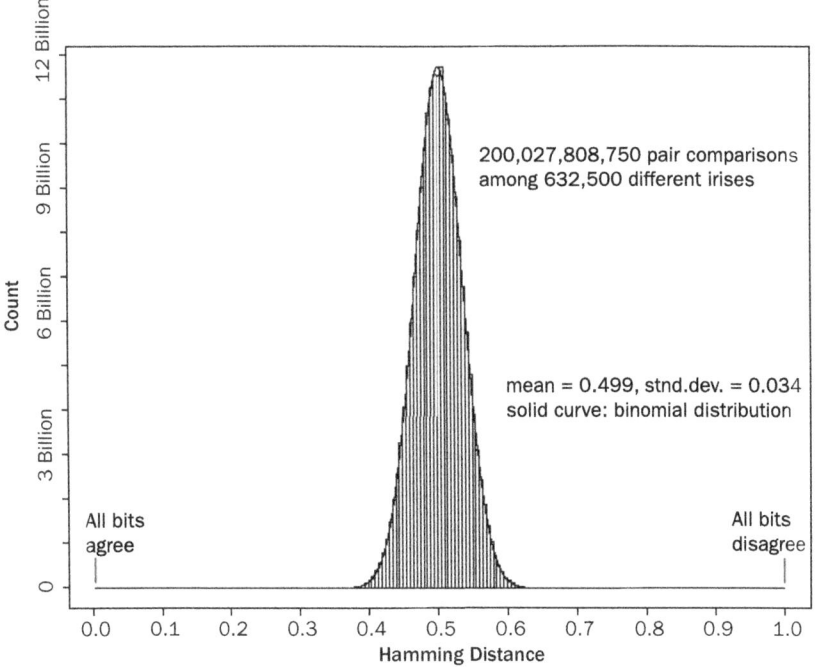

Figure 7.10 UAE study cross comparison results

Source: (Daugman, 2005)

Daugman took more than four weeks of computing and human power effort in Cambridge University Labs.

The study showed extreme accuracy of iris recognition and re-affirmed the management observations of the very high level of confidence in the collected IrisCodes™ and in all the subsequent matches that have taken place over the years in the UAE.

From a technical perspective, the study examined the distribution of similarity scores obtained from comparing different irises, and likewise the scores obtained from comparing different images of the same irises (see also Figure 7.10). These two distributions showed all cross comparison similarity scores obtained from making all possible pair comparisons between 632,500 different irises. The cumulative scores of

Table 7.4   Observed false match rates

HD Criterion	Observed False Match Rate
0.220	0 (theor: 1 in 5 x 10^{15})
0.225	0 (theor: 1 in 1 x 10^{15})
0.230	0 (theor: 1 in 3 x 10^{14})
0.235	0 (theor: 1 in 9 x 10^{13})
0.240	0 (theor: 1 in 3 x 10^{13})
0.245	0 (theor: 1 in 8 x 10^{12})
0.250	0 (theor: 1 in 2 x 10^{12})
0.255	0 (theor: 1 in 7 x 10^{11})
0.262	1 in 200 billion
0.267	1 in 50 billion
0.272	1 in 13 billion
0.277	1 in 2.7 billion
0.282	1 in 284 million
0.287	1 in 96 million
0.292	1 in 40 million
0.297	1 in 18 million
0.302	1 in 8 million
0.307	1 in 4 million
0.312	1 in 2 million
0.317	1 in 1 million

Source: Daugman (2005)

the distribution, up to various Hamming Distance thresholds, revealed the false match rates among the 200 billion iris comparisons if the identification decision policy used those thresholds.

As shown in Table 7.4, no such matches were found with Hamming Distances below about 0.260. The table has been extended down to 0.220 using equation (7) for extreme value samples of the binomial (plotted as the solid curve in Figure 7.10) to extrapolate the theoretically expected false match rates for such decision policies.

The performance test of the Daugman algorithms allowed conclusions to be drawn about the numerical decision policies that should be implemented in large-scale identity searches to ensure the absence of false matches and to calculate confidence levels.

The report stated the rule to be followed for decision policy threshold selection is to multiply the size of the enrolled database by the number of searches to be conducted against it in a given interval of time, and then to determine from the above table what Hamming Distance threshold will correspond to the risk level that is deemed to be acceptable. It is worth mentioning at this point that the study was carried out by the inventor of the algorithm himself, and is deemed important that further studies need to be carried out to validate the results of this study.

11. Lessons learned

Following are some lessons learned that were recorded during the implementation of the system.

As might be the case with any system implementation, lack of operator awareness and training may lead to resistance due to an inability to understand how the system works.

Periodic re-education and re-training programs were found effective in addressing this concern. Training programs played a significant role in promoting the appreciation of employee expectations and their willingness to accept the system as it eased the smooth implementation and operation of the system.

The system tested enrolment and recognition with people wearing eye glasses and contact lenses, and did not affect the accuracy or the speed of the system in almost all cases. However, dirty or scratched glasses in certain cases caused the inability of the system to recognise the iris.

In general, enrolees and travellers were asked to remove their glasses at the time of image acquisition. Some tests were carried also with joke contact lenses. The system was able to successfully detect such cases; the image was not recognised to carry a human pattern, and hence was rejected.

As indicated earlier, the UAE database represents 152 different nationalities. The system did not encounter any incident where it was unable to enrol people because of their gender, age, or racial differences, as was the case in other deployments in other countries.*

There were also those cases where some people used eye-drops to bypass the system. The eye-drops** were found to cause a temporary dilation of the pupil, meaning that the use of this substance will lead to a false rejection; the person is not found in the database.

In abnormal cases of dilation using some of the identified eye-drops, the ratio of the pupil radius to the iris radius exceeded 60 per cent. (see also Figure 7.11).

The UAE was a pioneer in solving this problem with the iris. The system was enhanced to reject any acquisition of irises where the ratio of the pupil to the iris is greater than 60 per cent and shows the ratio percentage on the screen for the operator.

In such cases a simple test was carried out using a pocket flashlight onto the eye. If no movement of the pupil is observed, then this person is most likely using an eye drop. A second check was required in a two-day timeframe after the effects had worn off.

Lighting and other environmental conditions were found to affect the acquisition process and therefore the functioning of the system. In many cases, calls received by the help desk reporting that 'the system is not working', were caused by insufficient lighting at those sites. This required the authorities to improve the enrolment and acquisition centres to ensure

---

* Other trial deployments in some countries faced problems such as enrolling Asian people and people with dark skin (black iris). The problem is most probably believed to be because of the type of cameras which were not able to detect the irises of those people. With 6.5 million travellers who used the system in the UAE to-date, there has been no single incident where the system failed to acquire an iris regardless of the gender, age, or racial differences.

** The most common use of this type of drops comes from ophthalmologists wishing to examine a patient's retina; the dilated pupil helps the physician to see better inside of the eye through the pupil's large opening. The effect of the eye-drops is temporary and the eye is back to normal in a day or two.

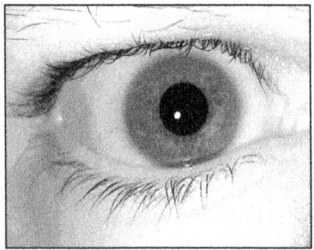 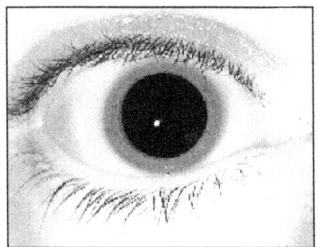

Normal dilation: pupil in a normal condition (ratio = 36 per cent)

Abnormal Dilation: The same pupil dilated by an eye drop called Atropine Sulphate (ratio = 71 per cent)

Figure 7.11 Pupil dilation

that the sources of bright white light (windows) were closed and that no sources of light were reflecting off the cornea and obscuring the iris.

The accuracy of the iris system since its implementation has been very impressive for the authorities. When it comes to enrolment, the system up to the time of writing this paper has had zero cases of FTE (failure to enrol), meaning that it was never unable to enrol a person for whatever reason.

As for false rejections meaning how many times the system failed to find an expelled person, thus allowing him into the country, the only measure to determine this factor was through the biographical information stored in the immigration system.

However, if the person changes this information, there will be no other way to determine this factor. This lies in the fact that the system is a negative application and should this happen, it would not be reported for obvious reasons (i.e., a former expellee will be happy that the system has failed to recognise him).

The local authorities invested in acquiring the most accurate cameras on the market and improving the acquisition environment as explained above. With this, the false rejection rate in the UAE system is indicated to be no more than 0.001 per cent according to the claim of the vendor. However, the system is designed to enrol both eyes of the person, where they are checked at recognition sites (e.g., airports and border points), so if one eye experiences a false rejection, the chance of the other experiencing the same is 0.01% x 0.01% = 0.0001% or one in a million.

12. Future applications

With such impressive results in the expellee project, the government is currently studying a proposed structure for building an integrated national iris repository for identification and verification purposes as depicted in Figure 7.12. Following are some key projects where the iris is considered to play a complementary role to support other biometrics for identification and authentication purposes, especially when rapid and real-time live detection is desired.

12.1 National ID project

This project is considered to be one of the most sophisticated technical projects in the Middle East, aiming to develop a modern identity management system that provides a secure and safe environment for the

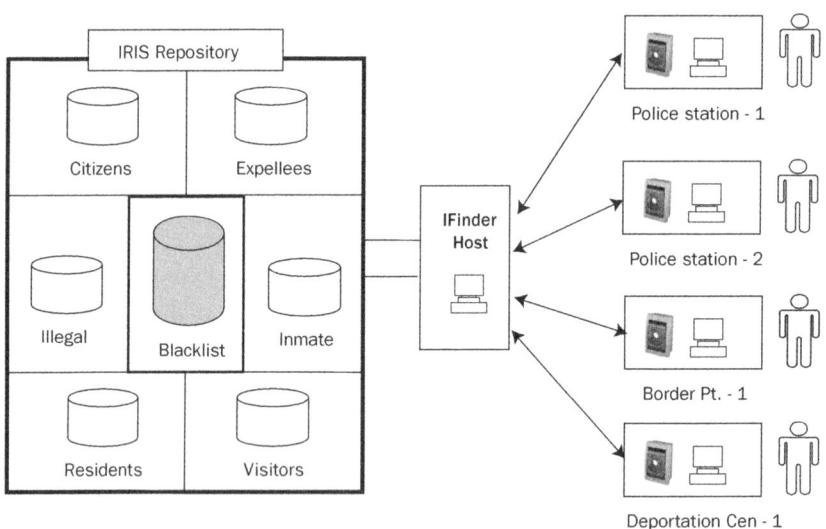

Figure 7.12 Proposed structure for a national iris repository

five million citizens and residents in the UAE. The government has plans to include the iris as a supporting biometric in the national ID project besides its current fingerprint technology based card.

12.2 e-Passports

In a step towards enhancing its border security control, the UAE government is in the process of launching a project to issue an RFID chip with biometric-enabled passports (e-passport). Iris recognition is being considered for inclusion in the pilot program that is planned for launch towards the end of next year.

12.3 Electronically operated gates (e-Gate)

In 2002, an electronic gate (e-Gate) project was launched at Dubai International Airport to allow frequent flyers fast access through immigration via electronically controlled gates. This fully automated passport control system replaces manual checks with proximity smart card and fingerprint technology to identify and clear registered passengers.

The government have recently launched a larger scale e-Gate project to cover all international airports in the UAE. Iris recognition is considered a viable option to support its current scheme and complement the currently used fingerprint biometric technology. The UAE and the UK Government (Immigration and Naturalisation Department – IND) are considering cooperating to use the same iris technology to enter the UK and UAE.

13. Conclusion

Technology is evolving at a very rapid speed. As for technological advances in terms of security, there are other groups of people who are always trying to get round such developments and exploit their weaknesses. Iris recognition, just as other biometric technologies, has been criticised for inefficiency and ineffectiveness. The UAE accepted the risk and pushed itself into a pioneering role and tested the system to become the world's largest iris database holder.

In its application in the UAE, iris recognition has proved a quick, reliable means of checking identity. In fact, the results presented in this study clearly show some very interesting facts about the system's performance. With 2.5 trillion comparisons performed to date on the system there has been a zero false match rate under a 0.262 Hamming Distance.

The full database search (1:N) is performed in less than three seconds. Having the largest iris database in the world with more than 840,751 iris records, the UAE government is satisfied with the results gained to date and is committed to taking part in the development of this technology as it is studying the incorporation of iris recognition in other high-tech projects such as electronic passport and national ID schemes.

Acknowledgment

The authors would like to thank Mr. Imad Malhas from IrisGuard Inc. for his feedback on this paper. They also would like to extend their gratitude to the editor and the reviewers of this article who provided feedback that improved the overall structure and quality of this paper.

Copyright

References

1. Daugman, J.G. (1994) 'Biometric Personal Identification System Based on Iris Analysis'. US patent 5,291,560, Patent and Trademark Office, Washington, D.C.
2. Daugman J.G. (2003) 'The importance of being random: Statistical principles of iris recognition'. *Pattern Recognition* 36: 279–291.
3. Daugman, J.G. (2005) 'The United Arab Emirates iris study: Results from 200 billion iris cross-comparisons'. University of Cambridge, UK.

4. Dillingham, G.L. (2002) 'Avaition Security. Registered Traveller Programme Policy and Implementation Issues'. General Accounting Office, USA. Available at: *http://frwebgate.access.gpo.gov/cgibin/useftp.cg- i?IPaddress=162.140.64.88&file-name=d03253.pdf&directory=/diskb/wais/-data/gao*

5. Flom, L. and Safir, A. (1987) 'Iris recognition system'. US Patent No. 4,641,349, Patent and Trademark Office, Washington, D.C.

6. Heath, D. (2001) 'An Overview of Biometrics Support in NetWare Through NMAS'. Novel, USA. Available at: *http://support.novell.com/tech-center/articles/ana20010701.html*

7. Liu S. and Silverman, M. (2001) 'A practical guide to biometric security technology'. *IT Professional*, 3 (1): 27–32.

8. Mansfield, T. (2001) 'Biometric Authentication in the real world'. National Physical Laboratory, UK. Available at: *http://www.npl.co.uk/s-cientific_software/publications/biometrics/ps-revho.pdf*

9. Mansfield, T. and Rejman-Greene, M. (2003) 'Feasibility Study on the Use of Biometrics in an Entitlement Scheme for UKPS, DVLA and the Home Office'. National Physical Laboratory, UK. Available at: *http://uk.sites-tat.com/homeoffice/homeoffice/s?docs2.feasibility_study031111_v2&ns_typ e=pdf*

9 781909 287518